Tesla's Magnifying Transmitter

recreating Tesla's dream

By E. W. van den Bergh

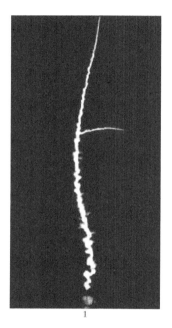

1   Characteristic sword-like discharge from a vacuum tube Tesla coil

Contents

Introduction..5
Part 1: Direct Evidence...9
   The TMT diagram..9
   What is electricity?...20
   Ether and "the medium"...24
   True wireless...30
   Conclusions of part 1...35
      What it a TMT?..35
      What does a TMT do?...35
      How does it do that?..36
Part 2: Diving Deeper...37
   The Tesla code..37
   Patent 1,113,716: Fountain..39
   A new source of energy..42
   The Tesla time line...49
      Tesla's autobiography...50
      Increasing Human Energy...55
   Combining the pieces..63
      Sir William Crookes...64
      The open vacuum tube..66
      The "extra coil"..67
      Elements 4 and 5..69
   Cosmic rays...70
      2nd Law of Thermodynamics..70
      Energy from cosmic rays...73
      One story told 5 times..74
   Conclusions of part 2...79
      The Tesla code..79
      New source of energy..79
      Cosmic rays...79
Part 3: Experimental Evidence...81
   Natural evidence..81
   Ring experiment...83
   Resonance measurement...83
   IR thermometer..84
   Single terminal tube..85
   Earth resonance...85

- Charge increase ... 86
- Conclusions of part 3 ... 88
  - True or not? ... 88
- Appendices ... 89
  - Patent application 213,055 ... 89
  - Tesla's dynamic theory of gravity ... 93
  - Capturing cosmic rays ... 94
- Help bring back the Magnifying Transmitter ... 97

Introduction

Around 1900 Nikola Tesla had an enormous wooden tower built on Long Island, New York on a property that he called Wardenclyffe.

This tower housed what Tesla called his Magnifying Transmitter and was the first part of his envisioned *"World System"*. This *"World System"* would provide among other things worldwide wireless power, telegraphy, telephony and navigation. To start such an expensive project one would assume either Tesla was very certain of success or he was the "mad scientist" that he is often portrayed as.

As the inventor of the induction motor, our complete AC distribution network and many other things, Tesla appears to be quite an talented man. At least he was so up until 1889.

So what happened after that date?

Quoting Matthew Inman, known for "The Oatmeal", not for his knowledge of Nikola Tesla:

"Regarding Tesla's mental health, his cheese had pretty much slid off his cracker by the time he was 30 years old. In 1905 when Einstein's theory of relativity was gaining momentum, Tesla was basically clinically insane by then."

30 years old? That was in 1886. He gave some very well received lectures between 1891-93, and demonstrated the first radio controlled boat in 1898. Quite impressive for someone "without cheese on his cracker".

After studying Tesla's work, his articles, his notes, his letters and his patents one can only conclude that we are dealing with meticulous work of an absolute genius. Because his work is so thorough it is difficult to assume that he was on the wrong track.

So assuming then that he was a genius and he was really onto something, then what was it?

That is the question that I set out to answer over a decade ago and was the trigger for my research.

I started reading what others had said but quickly concluded that that was not the right approach, I had to turn to the only reliable source being Tesla's own writings.

We are lucky that so much has survived and is freely available on the internet, while even more can be purchased at the Tesla Museum[2]. With such a rich source of sound information there is really no need to be led astray by others. Although there are two other sources that I believe are worth mentioning. The first is Margaret Cheney with her books "Tesla: Man Out of Time" and "Tesla: Master of Lightning". The second is Marc J. Seifer with his book "Wizard". While Tesla's own writings are my primary source, these two provide good background information.

To find out what Tesla called his Magnifying Transmitter, what it does and how it does so, we have 3 types of source material;

1 – direct evidence

2 – circumstantial evidence

3 – a hidden "code" in Tesla's work after 1895

The first are direct statements of Tesla about his TMT[3], such as my TMT does ..., or my TMT is composed of ... etc.

The second are indirect statements or images from which one can draw certain conclusions regarding the TMT and also information obtained by comparing various articles and time-lines.

The third is something I came across while studying his articles and letters. The most striking example is found in his article "the Problem of Increasing Human Energy", published in the Century Illustrated Magazine on June 1900.

This appears to be a very strange, vague and philosophical article about many other things *outside* of Tesla's field of expertise. It thus stands in sharp contrast with his other work which is very precise, complete and to the point. But reading it over and over again certain

---

2   Most worthwhile are "From Colorado Springs to Long Island" and "The Unresolved Patents of Nikola Tesla"

3   For the sake of brevity I will use the abbreviation TMT (Tesla's Magnifying Transmitter) from here on.

patterns emerge, there seems to be a message hidden in this text. I believe I have been able to recover the greater part of this message which covers Tesla's theories and work from 1889 onwards.

In this book I will take you through these sources and piece the pieces together.

My research has not been limited to a purely theoretical study. As much as my resources and other practical considerations allowed I have been repeating the experiments that Tesla described so that I could see first-hand what Tesla must have seen and thus align my line of reasoning with his.

One needs to realise that in order to understand Tesla we must go back in time to his days, over a century ago. Many terms that bring clear and well-defined images to our minds, were not so well-defined at his time. For example when we hear the word atom, we immediately see a picture of a nucleus made up of protons and neutrons with a cloud of electrons swirling around. This is not at all what Tesla would have thought of when he used that word. Tesla would imagine an indivisible elementary particle that makes up a certain material.

Tesla did not at all agree with Einstein's theory and even less with quantum mechanics. Tesla believed in an ether, a fluid incompressible medium that fills up all space, and more than that, in a gaseous medium submerged in the ether that formed the root cause of electrical phenomena.

Thus our education strongly interferes with trying to understand Tesla's theories. We have to open our minds to a completely different interpretation of experimental results.

Finally, as I am not a native English speaker I hope you will excuse me for my poor English. But this is not intended to be a literary masterpiece and I think you will get the gist of what I want to convey.

Having said all this, let's not push it off any longer and dive right into it.

Part 1: Direct Evidence

The TMT diagram

I would have loved to mention the original name of the TMT here: "Self-Regenerative Resonant Transformer", but I read this in a publication not written by Tesla himself and I can not find this term in any part of Tesla's original work.

Since this name is a very good description of the TMT as we will see later on, I do believe that Tesla may indeed have used this name, I just have not found any evidence.

Talking of which, there is not really much "direct evidence", but here we go.

1904, March 5th

> **The Transmission of Electrical Energy Without Wires**
>
> *With these[4] stupendous possibilities in sight, and the experimental evidence before me that their realization was henceforth merely a question of expert knowledge, patience and skill, I attacked vigorously the development of my magnifying transmitter, now, however, not so much with the original intention of producing one of great power, as with the object of learning how to construct the best one. This is, essentially, a circuit of very high self-induction and small resistance which in its arrangement, mode of excitation and action, may be said to be the diametrical opposite of a transmitting circuit typical of telegraphy by Hertzian or electromagnetic radiations. It is difficult to form an adequate idea of the marvellous power of this unique appliance, by the aid of which the globe will be transformed. The electromagnetic radiations being reduced to an insignificant quantity, and proper conditions of resonance maintained, the circuit acts like an immense pendulum, storing indefinitely the*

---
4 Refers to the transmission of energy through the Earth, using Earth resonance

energy of the primary exciting impulses and impressing upon the earth and its conducting atmosphere uniform harmonic oscillations of intensities which, as actual tests have shown, may be pushed so far as to surpass those attained in the natural displays of static electricity.

Key points:
- very high inductance and small resistance
- opposite of traditional transmitting circuit
- acts like immense pendulum, accumulating energy of primary exciting impulses
- impresses upon the earth and its conducting atmosphere uniform harmonic oscillations

1904, March 27th

Cloudborn Electric Wavelets To Encircle The Globe

But in his article he announces that he will transmit from the tower an electric wave of a total maximum activity of ten million horse power. This, he says, will be possible with a plant of but 100 horse power, by the use of a magnifying transmitter of his own invention and certain artifices which he promises to make known in due course.

Note, this is not 100% *direct* evidence because in this quote the author says he quotes Tesla. So we need to trust this author. I still wanted to keep this one here as the statement seems quite remarkable, producing 10,000,000 HP using only 100.

This requires some closer examination.

First, power is energy per unit of time. So if I use 1 Joule in 1 second, that is 1 Watt, but if I accumulate that energy and then release it in 1 micro second, that same energy becomes 1,000,000 Watt. Tesla does something very similar with his Tesla coil, he charges a capacitor at a slow and steady power rate and then using a spark-gap that energy is released in a very short time interval creating a very powerful pulse.

Second, there is a mix-up here in the usage of the word power. The

100 HP refers to a constant expenditure of energy, while the 10,000,000 HP refers to the power built-up in a wave. The latter being the energy in the wave divided by its period, i.e. a fixed time interval.

So for example, if we have a 10 Hz wave and I add 100 HP to it for 1 second, the wave will have a power of 1,000 HP. Now suppose through natural occurring losses the wave loses its energy at a 0.1% per second rate, then the energy (and thus the power) in the wave will build-up until the losses equal the added power, which happens when there is 100,000 HP in the wave.

Still to go from 100 to 10,000,000 HP implies either extremely small losses or added power from another source. The latter being the more likely, that is, if we could trust this author quoting Tesla.

1905, January 7th

The Transmission of Electrical Energy Without Wires as a Means for Furthering Peace

That electrical energy can be economically transmitted without wires to any terrestrial distance, I have unmistakably established in numerous observations, experiments and measurements, qualitative and quantitative. These have demonstrated that is practicable to distribute power from a central plant in unlimited amounts, with a loss not exceeding a small fraction of one per cent, in the transmission, even to the greatest distance, twelve thousand miles—to the opposite end of the globe. This seemingly impossible feat can now be readily performed by any electrician familiar with the design and construction of my "high-potential magnifying transmitter," the most marvellous electrical apparatus of which I have knowledge, enabling the production of effects of unlimited intensities in the earth and its ambient atmosphere. It is, essentially, a freely vibrating secondary circuit of definite length, very high self-induction and small resistance, which has one of its terminals in intimate direct or inductive connection with the ground and the other with an elevated conductor, and upon which the electrical oscillations of a primary or exciting circuit are impressed under conditions of resonance. To give an idea of the capabilities of this wonderful

> *appliance, I may state that I have obtained, by its means, spark discharges extending through more than one hundred feet and carrying currents of one thousand amperes, electromotive forces approximating twenty million volts, chemically active streamers covering areas of several thousand square feet, and electrical disturbances in the natural media surpassing those caused by lightning, in intensity.*

Key points:
- effects of *unlimited* intensities
- in the Earth and its atmosphere
- a freely vibrating secondary circuit of definite length,
- very high self-induction and small resistance
- one of its terminals connected to the ground and the other to an elevated conductor
- upon which the oscillations of a primary or exciting circuit are impressed under conditions of resonance
- 1,000 Amps and 20,000,000 Volts (being 20 GW or 26.8 million HP)

Within these key points we recognize a few that were mentioned earlier.

1907, April 21st and May 3rd

Tesla's Tidal Wave to Make War impossible

> *Electricity can be stored in the form of explosive energy of a violence against which the detonation of cordite is but a breath. With a magnifying transmitter as diagrammatically illustrated, rates of 25,000,000 H.P. have already been obtained. A similar and much improved machine, now under construction, will make it possible to attain maximum explosive rates of over 800,000,000 H.P., twenty times the performance of the Dreadnought's broadside of eight 12 in. guns simultaneously fired.*

Note: I have reprints of two appearances of this article, one in the

"New York World" of April 21st, and one in the "English Mechanic and World of Science" of May 3rd. Both articles do not show an illustration of a TMT. For some reason the editors have chosen to leave that out. (?)

Key points:

- rates of 25,000,000 HP have been obtained
- Wardenclyffe will be able to do 800,000,000 HP (597 GW)

The first number we already calculated from the data provided in the previous article. While 597 GW certainly is a lot, it is not *unlimited* as stated earlier.

1916

Pre-Hearing Interview.

This coil, which I have subsequently shown in my patents Nos. 645,576 and 649,621, in the form of a spiral, was, as you see, [earlier] in the form of a cone. The idea was to put the coil, with reference to the primary, in an inductive connection which was not close—we call it now a loose coupling—but free to permit a great resonant rise. That was the first single step, as I say, toward the evolution of an invention which I have called my "magnifying transmitter." That means, a circuit connected to ground and to the antenna, of a tremendous electromagnetic momentum and small damping factor, with all the conditions so determined that an immense accumulation of electrical energy can take place.

Key points:

- a circuit connected to ground and to the antenna
- of a tremendous electromagnetic momentum and small damping factor
- an immense accumulation of electrical energy can take place

All of these points we have seen before, and therefore this seems of little relevance but this piece provides something more, something very essential. So let's read on.

Counsel

Mr. Tesla, at that point, what did you mean by electro-magnetic momentum?

Tesla

I mean that you have to have in the circuit, inertia. You have to have a large self-inductance in order that you may accomplish two things: First, a comparatively low frequency, which will reduce the radiation of the electromagnetic waves to a comparatively small value, and second, a great resonant effect. That is not possible in an antenna, for instance, of large capacity and small self-inductance. A large capacity and small self-inductance is the poorest kind of circuit which can be constructed; it gives a very small resonant effect. That was the reason why in my experiments in Colorado the energies were 1,000 times greater than in the present antennae.

Counsel

You say the energy was 1,000 times greater. Do you mean that the voltage was increased, or the current, or both?

Tesla

Yes [both]. To be more explicit, I take a very large self-inductance and a comparatively small capacity, which I have constructed in a certain way so that the electricity cannot leak out. I thus obtain a low frequency; but, as you know, the electromagnetic radiation is proportionate to the square root of the capacity divided by the self-induction. I do not permit the energy to go out; I accumulate in that circuit a tremendous energy. When the high potential is attained, if I want to give off electromagnetic waves, I do so, but I prefer to reduce those waves in quantity and pass a current into the earth, because electromagnetic wave energy is not recoverable while that [earth] current is entirely recoverable, being the energy stored in an elastic system.

Counsel

What elastic system do you refer to?

Tesla

I mean this: If you pass a current into a circuit with large self-induction, and no radiation takes place, and you have a low resistance, there is no possibility of this energy getting out into space; therefore, the impressed impulses accumulate.

What have we learned?

- electro-magnetic momentum refers to self-inductance.
- The secondary circuit contains a very large self-inductance and a comparatively small capacity.
- The Earth current is stored in an elastic system.
- This elastic system is the Earth, having a huge self-inductance and small resistance.

This solves a very crucial issue, because Tesla refers to a very large inductor in all previous articles, but we haven't seen such a huge coil in Colorado Springs nor in Wardenclyffe. Now we know what the secondary circuit is; the cupola of the tower (comparatively small capacity), the Earth (very large self-inductance) and a conductor connecting these two.

This matches the diagrams provided by Leland Anderson to a magazine called "the Electric Spacecraft", which they published in an article called "Rare Notes from Tesla on Wardenclyffe"[5] in 1997.

One of these can also be found in a publication of the Tesla Museum in Belgrade "From Colorado Springs to Long Island", together with a number of variations thereon. With this knowledge one can also recognize its development in diagrams from "the Colorado Springs Notes".

So now we have established the diagram of the TMT with absolute certainty, and consequently the fact that patent 1,119,732 which people commonly refer to as a TMT is *not* what Tesla would have called a TMT.

---

5 I shall abbreviate that to "Rare Notes" from here on

1919

My Inventions[6]

I have been asked by the ELECTRICAL EXPERIMENTER to be quite explicit on this subject so that my young friends among the readers of the magazine will clearly understand the construction and operation of my "Magnifying Transmitter" and the purposes for which it is intended. Well, then, in the first place, it is a resonant transformer with a secondary in which the parts, charged to a high potential, are of considerable area and arranged in space along ideal enveloping surfaces of very large radii of curvature, and at proper distances from one another thereby insuring a small electric surface density everywhere so that no leak can occur even if the conductor is bare. It is suitable for any frequency, from a few to many thousands of cycles per second, and can be used in the production of currents of tremendous volume and moderate pressure, or of smaller amperage and immense electromotive force. The maximum electric tension is merely dependent on the curvature of the surfaces on which the charged elements are situated and the area of the latter.

Judging from my past experience, as much as 100,000,000 volts are perfectly practicable. On the other hand currents of many thousands of amperes may be obtained in the antenna. A plant of but very moderate dimensions is required for such performances. Theoretically, a terminal of less than 90 feet in diameter is sufficient to develop an electromotive force of that magnitude while for antenna currents of from 2,000-4,000 amperes at the usual frequencies it need not be larger than 30 feet in diameter.

In a more restricted meaning this wireless transmitter is one in which the Hertz-wave radiation is an entirely negligible quantity as compared with the whole energy, under which condition the damping factor is extremely small and an enormous charge is stored in the elevated capacity. Such a circuit may then be excited with impulses of any kind, even of low frequency and it will yield sinusoidal and continuous

---

6 Tesla's autobiography published in "The Electrical Experimenter" from February to October, 1919

oscillations like those of an alternator.

Taken in the narrowest significance of the term, however, it is a resonant transformer which, besides possessing these qualities, is accurately proportioned to fit the globe and its electrical constants and properties, by virtue of which design it becomes highly efficient and effective in the wireless transmission of energy. Distance is then absolutely eliminated, there being no diminution in the intensity of the transmitted impulses. It is even possible to make the actions increase with the distance from the plant according to an exact mathematical law.

---/-/---

My belief is firm in a law of compensation. The true rewards are ever in proportion to the labor and sacrifices made. This is one of the reasons why I feel certain that of all my inventions, the Magnifying Transmitter will prove most important and valuable to future generations.

Having read all the previous articles it is very clear what Tesla is referring to in this article.

Key points:

- it is a resonant transformer plus …
- … a secondary in which the parts, charged to a high potential, are of considerable area and arranged in space along ideal enveloping surfaces of very large radii of curvature, and at proper distances from one another thereby insuring a small electric surface density everywhere so that no leak can occur even if the conductor is bare
- suitable for any frequency
- 100,000,000 volts are perfectly practicable
- currents of many thousands of amperes may be obtained
- A plant of but very moderate dimensions is required for such performances
- the Hertz-wave radiation is an entirely negligible quantity

- the damping factor is extremely small
- an enormous charge is stored in the elevated capacity
- circuit may then be excited with impulses of any kind
- it will yield sinusoidal and continuous oscillations
- it is a resonant transformer which is accurately proportioned to fit the globe and its electrical constants and properties

Note the mention of "surfaces of *very large* radii of curvature", yet a plant of but *very moderate dimensions* is required. Now we understand that the very large radii refers to the secondary circuit which includes the Earth itself, and thus the plant can be of very moderate dimensions.

Although most points have been made earlier we also learn a few new things, mainly about the primary or exciting circuit. Namely that this is a resonant transformer that may operate on *any* frequency and give impulses of *any* kind.

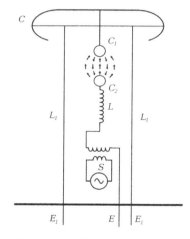

Here we have the result so far. S, L and C_2 form a resonant transformer which constitutes the primary circuit. In this example the resonant transformer is equipped with a 3rd, so called "extra" coil, but this is not a necessity. Tesla's notes also show variations without this 3rd coil.

The secondary circuit is formed by C, L_1 and the Earth.

The resonance frequency of C and L_1 may differ from that of the primary circuit[7]. Tesla played with the idea to make multiple connections between C and the Earth each with a different induction, thus creating a system that provided more than one base frequency. Tesla needed more frequencies for his "art of individualization" which would make it possible to send messages to specific receivers.

One may ask if we have proof that *this* was what Tesla was building in Wardenclyffe and not the image drawn in patent 1,119,732 that so

---

7   Tesla states in the "Rare Notes"

many people believe to be a TMT.

Well, to those I can answer: Yes, we actually have!.

In the "Wardenclyffe Foreclosure Appeal Proceedings" of 1922 at line 737 and 738 we read:

> *This big ball on top of the tower, you could not tell what it was made out of, whether it was brass or steel, as the ends of <u>the wires where it had been grounded</u>[8] had rusted out and blown away, and there was a thousand and one little wires sticking out in every direction, so you could not see what it was made up of.*

In other words the cupola had a ground connection! This is in contradiction with the patent and matches our diagram.

We are only a few pages into this book and we already know what a TMT is, or do we?

It is clear that Tesla intended this machine to send power and messages across the world, but how and were there perhaps other things that this machine could do that he was less vocal about?

To answer these questions we must first take a step back and try to find out how Tesla viewed electricity.

---

8   Emphasis (underline) added.

What is electricity?

To answer this question, or rather, to have Tesla answer this question we turn to his lecture delivered before the American Institute of Electrical Engineers, Columbia College, N.Y. on May 20th, 1891.

Experiments with Alternate Currents of Very High Frequency and Their Application to Methods of Artificial Illumination

First, we naturally inquire, What is electricity, and is there such a thing as electricity? In interpreting electric phenomena: we may speak of electricity or of an electric condition, state or effect. If we speak of electric effects we must distinguish two such effects, opposite in character and neutralizing each other, as observation shows that two such opposite effects exist. This is unavoidable, for in a medium of the properties of ether, we cannot possibly exert a strain, or produce a displacement or motion of any kind, without causing in the surrounding medium an equivalent and opposite effect. But if we speak of electricity, meaning a thing, we must, I think, abandon the idea of two electricities, as the existence of two such things is highly improbable. For how can we imagine that there should be two things, equivalent in amount, alike in their properties, but of opposite character, both clinging to matter, both attracting and completely neutralizing each other? Such an assumption, though suggested by many phenomena, though most convenient for explaining them, has little to commend it. If there is such a thing as electricity, there can be only one such thing, and; excess and want of that one thing, possibly; but more probably its condition determines the positive and negative character. The old theory of Franklin, though falling short in some respects; is, from a certain point of view, after all, the most plausible one. Still, in spite of this, the theory of the two electricities is generally accepted, as it apparently explains electric phenomena in a more satisfactory manner. But a theory which better explains the facts is not necessarily true. Ingenious minds will invent theories to suit observation, and almost every independent thinker has his own views on the subject.

It is not with the object of advancing an opinion; but with the desire of acquainting you better with some of the results, which I will describe, to show you the reasoning I have followed, the departures I have made—that I venture to express, in a few words, the views and convictions which have led me to these results.

I adhere to the idea that there is a thing which we have been in the habit of calling electricity. The question is, What is that thing? or, What, of all things, the existence of which we know, have we the best reason to call electricity? We know that it acts like an incompressible fluid; that there must be a constant quantity of it in nature; that it can be neither produced nor destroyed; and, what is more important, the electro-magnetic theory of light and all facts observed teach us that electric and ether phenomena are identical. The idea at once suggests itself, therefore, that electricity might be called ether. In fact, this view has in a certain sense been advanced by Dr. Lodge. His interesting work has been read by everyone and many have been convinced by his arguments. His great ability and the interesting nature of the subject, keep the reader spellbound; but when the impressions fade, one realizes that he has to deal only with ingenious explanations. I must confess, that I cannot believe in two electricities, much less in a doubly-constituted ether. The puzzling behaviour of the ether as a solid to waves of light and heat, and as a fluid to the motion of bodies through it, is certainly explained in the most natural and satisfactory manner by assuming it to be in motion, as Sir William Thomson has suggested; but regardless of this, there is nothing which would enable us to conclude with certainty that, while a fluid is not capable of transmitting transverse vibrations of a few hundred or thousand per second, it might not be capable of transmitting such vibrations when they range into hundreds of million millions per second. Nor can anyone prove that there are transverse ether waves emitted from an alternate current machine, giving a small number of alternations per second; to such slow disturbances, the ether, if at rest, may behave as a true fluid.

Returning to the subject, and bearing in mind that the

existence of two electricities is, to say the least, highly improbable, we must remember, that we have no evidence of electricity, nor can we hope to get it, unless gross matter is present. Electricity, therefore, cannot be called ether in the broad sense of the term; but nothing would seem to stand in the way of calling electricity ether associated with matter, or bound to it; or, in other words, that the so-called static charge of the molecule is ether associated in some way with the molecule. Looking at it in that light, we would be justified in saying, that electricity is concerned in all molecular actions.

Now, precisely what the ether surrounding the molecules is, wherein it differs from ether in general, can only be conjectured. It cannot differ in density, ether being incompressible; it must, therefore, be under some strain or in motion, and the latter is the most probable. To understand its functions, it would be necessary to have an exact idea of the physical construction of matter, of which, of course, we can only form a mental picture.

But of all the views on nature, the one which assumes one matter and one force, and a perfect uniformity throughout, is the most scientific and most likely to be true. An infinitesimal world, with the molecules and their atoms spinning and moving in orbits, in much the same manner as celestial bodies, carrying with them and probably spinning with them ether, or in other words; carrying with them static charges, seems to my mind the most probable view, and one which, in a plausible manner, accounts for most of the phenomena observed. The spinning of the molecules and their ether sets up the ether tensions or electrostatic strains; the equalization of ether tensions sets up ether motions or electric currents, and the orbital movements produce the effects of electro and permanent magnetism.

About fifteen, years ago, Prof. Rowland demonstrated a most interesting and important fact; namely, that a static charge carried around produces the effects of an electric current. Leaving out of consideration the precise nature of the mechanism, which produces the attraction and repulsion of currents, and conceiving the electrostatically charged molecules in motion, this experimental fact gives us a fair idea of magnetism. We can conceive lines or tubes of force which physically exist, being formed of rows of directed moving molecules; we can see that these lines must be closed, that they must tend to shorten and expand, etc. It likewise explains in a reasonable way, the most puzzling phenomenon of all, permanent magnetism, and, in general, has all the beauties of the Ampere theory without possessing the vital defect of the same, namely, the assumption of molecular currents. Without enlarging further upon the subject, I would say, that I look upon all electrostatic, current and magnetic phenomena as being due to electrostatic molecular forces.

Key points:

- there is only *one* electricity causing both positive and negative charges
- it acts like an incompressible fluid
- there must be a constant quantity of it in nature
- electricity can be seen as ether associated with matter, or bound to it
- this ether is probably set in motion by the atoms that it is bound to
- an infinitesimal world, with the molecules and their atoms spinning and moving in orbits, in much the same manner as celestial bodies, carrying with them and probably spinning with them ether, or in other words; carrying with them static charges, seems to my mind the most probable view
- the spinning of the molecules and their ether sets up the ether tensions or electrostatic strains; the equalization of ether tensions sets up ether motions or electric currents, and the orbital movements produce the effects of electro and permanent magnetism
- we can conceive lines or tubes of magnetic force which physically exist, being formed of rows of directed moving molecules; we can see that these lines must be closed, that they must tend to shorten and expand

24

The first point is by far the most important to keep in mind when reading Tesla's work for today when we think of electricity we immediately think of electrons, static or moving charges. We do not ask ourselves what causes these charges to exist.

Electrons *can not* be the root cause of electricity as they are only concerned in negative charges. They only paint half the picture.

Ether and "the medium"

Tesla firmly believed in the ether, an concept abandoned by modern science. Reading his work we also come across the term "the medium" and if we are not careful it is easy to assume that these are the same. This is however not the case.

Tesla's ether is very similar to vedic akasha; a fluid incompressible medium that fills up all space. "The medium" is a term Tesla employs when talking about the root cause of electric effects. The latter looks somewhat like the photon gas or electron gas that we know today, and I suggest to just call it "electricity".

Let's look at the relevant quotes.

1891, May 20th

> **Experiments with Alternate Currents of Very High Frequency and Their Application to Methods of Artificial Illumination**

image courtesy of the Tesla Collection

> *Nature has stored up in the universe infinite energy. The eternal recipient and transmitter of this infinite energy is the ether. The recognition of the existence of ether, and of the functions it performs, is one of the most important results of modern scientific research. The mere abandoning of the idea of action at a distance, the assumption of a medium pervading all space and connecting all gross matter, has freed the minds of thinkers of an ever present doubt, and,*

by opening a new horizon—new and unforeseen possibilities—has given fresh interest to phenomena with which we are familiar of old. It has been a great step towards the understanding of the forces of nature and their multifold manifestations to our senses.

In this first quote we read that Tesla was a firm believer in the ether. And the last quote here will show that Tesla remained so until he died in 1943. But something changed over the years. Tesla found that this ether does not perform many of the functions that were attributed to it. Here is a second quote from the same article:

What is electricity, and what is magnetism? These questions have been asked again and again. The most able intellects have ceaselessly wrestled with the problem; still the question has not as yet been fully answered. But while we cannot even to-day state what these singular forces are, we have made good headway towards the solution of the problem. We are now confident that electric and magnetic phenomena are attributable to ether, and we are perhaps justified in saying that the effects of static electricity are effects of ether under strain, and those of dynamic electricity and electro-magnetism effects of ether in motion. But this still leaves the question, as to what electricity and magnetism are, unanswered.

This continues into the quote from our previous chapter starting with "First, we naturally inquire, what is electricity, ...".

Tesla's view here is not yet matured as we can read about 1 year later in his lecture of February 3rd, 1892.

Experiments with Alternate Currents of High Potential and High Frequency

What impresses the investigator most in the course of these experiences is the behaviour of gases when subjected to great rapidly alternating electrostatic stresses. But he must remain in doubt as to whether the effects observed are due wholly to the molecules, or atoms, of the gas which chemical analysis discloses to us, or whether there enters into play another medium of a gaseous nature, comprising atoms, or molecules, immersed in a fluid pervading the space. Such a medium, surely must exist, and I am convinced that, for instance, even if

image courtesy of the Tesla Collection

air were absent, the surface and neighbourhood of a body in space would be heated by rapidly alternating the potential of the body; but no such heating of the surface or neighbourhood could occur if all free atoms were removed and only a homogeneous, incompressible, and elastic fluid—such as ether is supposed to be—would remain, for then there would be no impacts, no collisions. In such a case, as far as the body itself is concerned, only frictional losses in the inside could occur.

Key points:
- there enters into play another medium of a gaseous nature, comprising atoms, or molecules, immersed in a fluid pervading the space. Such a medium, surely <u>must</u> exist.

Again 1 year later in his lecture of February 24th, 1893:

On Light and Other High Frequency Phenomena

image courtesy of the Tesla Collection

It is certainly more in accordance with many phenomena observed with high frequency currents to hold that all space is pervaded with free atoms, rather than to assume that it is devoid of these, and dark and cold, for so it must be, if filled with a continuous medium, since in such there can be neither heat nor light. Is then energy transmitted by independent carriers or by the vibration of a continuous medium? This important question is by no means as yet positively answered. But most of the effects which are here considered, especially the light effects, incandescence, or phosphorescence, involve the presence of free atoms and would be impossible without these.

This merely enforces the key point of the previous quote.

1905, January 7th

The Transmission of Electrical Energy Without Wires as a Means for Furthering Peace

Our accepted estimates of the duration of natural metamorphoses, or changes in general, have been thrown in doubt of late. The very foundations of science have been shaken. We can no longer believe in the Maxwellian hypothesis of transversal ether-undulations of electrical vibrations, this most important field of human endeavour, particularly in the advancement of philanthropy and peace, was in no small measure retarded by that fascinating illusion, which I since long hoped to dispel. I have noted with satisfaction the first signs of a change of scientific opinion. The brilliant discovery of the exceptionally "radio-active" substances, radium and polonium, by Mrs. Sklodowska Curie, has likewise afforded me much personal gratification, being an eclatant confirmation of my early experimental demonstrations, of electrified radian streams of primary matter or corpuscular emanations (Electrical Review, New York, 1896-1897), which were then received with incredulity. They have awakened us from the poetical dream of an intangible conveyor of energy, weightless, structureless ether, to the plain, palpable reality of a ponderous medium of coarse particles, or bodily carriers of force. They have led us to a radically new interpretation of the changes and transformations we perceive. Enlightened by this recognition, we cannot say the sun is hot, the moon is cold, the star is bright, for all these might be purely electrical phenomena. If this be the case, then even our conceptions of time and space may have to be modified.

If this were the last article we would have to conclude that Tesla no longer believed in the ether. But this does not appear to be true. Tesla did no longer believe in the ether as a carrier of electric and magnetic effects. But he still believed in an ether which is the ultimate building block of matter as defined by Lord Kelvin and long before him in the

Vedas. Electric and magnetic effects are caused by what Tesla now refers to as "the medium", a gaseous medium consisting of minute particles. Probably of a much smaller scale than that of electrons.

May 1919

True Wireless

In 1900, however, after I had evolved a wireless transmitter which enabled me to obtain electro-magnetic activities of many millions of horse-power, I made a last desperate attempt to prove that the disturbances emanating from the oscillator were ether vibrations akin to those of light, but met again with utter failure. For more than eighteen years I have been reading treatises, reports of scientific transactions, and articles on Hertz-wave telegraphy, to keep myself informed, but they have always imprest me like works of fiction.

1929, September 22nd

Nikola Tesla Tells of New Radio Theories

When Dr. Heinrich Hertz undertook his experiments from 1887 to 1889 his object was to demonstrate a theory postulating a medium filling all space, called the ether, which was structureless, of inconceivable tenuity and yet solid and possessed of rigidity incomparably greater than that of the hardest steel. He obtained certain results and the whole world acclaimed them as an experimental verification of that cherished theory. But in reality what he observed tended to prove just its fallacy.

I had maintained for many years before that such a medium as supposed could not exist, and that we must rather accept the view that all space is filled with a gaseous substance. On repeating the Hertz experiments with much improved and very powerful apparatus, I satisfied myself that what he had observed was nothing else but effects of longitudinal waves in a gaseous medium, that is to say, waves, propagated by alternate compression and expansion. He had observed waves in the ether much of the nature of sound waves in the air.

Up to 1896, however, I did not succeed in obtaining a positive experimental proof of the existence of such a medium. But in that year I brought out a new form of vacuum tube capable of being charged to any desired potential, and operated it with effective pressures of about 4,000,000 volts. I produced cathodic and other rays of transcending intensity. The effects, according to my view, were due to minute particles of matter carrying enormous electrical charges, which, for want of a better name, I designated as matter not further decomposable. Subsequently those particles were called electrons.[9]

One of the first striking observations made with my tubes was that a purplish glow for several feet around the end of the tube was formed, and I readily ascertained that it was due to the escape of the charges of the particles as soon as they passed out into the air; for it was only in a nearly perfect vacuum that these charges could be confined to them. The coronal discharge proved that there must be a medium besides air in the space, composed of particles immeasurably smaller than those of air, as otherwise such a discharge would not be possible. On further investigation I found that this gas was so light that a volume equal to that of the earth would weigh only about one- twentieth of a pound.[10]

The velocity of any sound wave depends on a certain ratio between elasticity and density, and for this ether or universal gas the ratio is 800,000,000,000 times greater than for air. This means that the velocity of the sound waves propagated through the ether is about 300,000 times greater than that of the sound waves in air, which travel at approximately 1,085 feet a second. Consequently the speed in ether is 900,000 x 1,085 feet, or 186,000 miles, and that is the speed of light.

Key points:

Although Tesla previously stated that electricity acts as an incompressible medium we read here that Tesla talks about "longitudinal waves in a gaseous medium". This requires that this medium is compressible, just like any other gas. We shall see later on that Tesla uses this possibility, so he must be fully aware of it.

---

9   Note that Tesla's view of electrons is different from ours
10 $2.09374 \cdot 10^{-26}$ g/ml

1930, July 6th

Man's Greatest Achievement

Long ago he recognized that all perceptible matter comes from a primary substance, of a tenuity beyond conception and filling all space - the Akasa or luminiferous ether - which is acted upon by the life-giving Prana or creative force, calling into existence, in never ending cycles, all things and phenomena. The primary substance, thrown into infinitesimal whirls of prodigious velocity, becomes gross matter; the force subsiding, the motion ceases and matter disappears, reverting to the primary substance.

There are a few more articles in which Tesla talks about this matter but these do not add anything to what we already have.

I think this last article is relevant because it shows that Tesla in 1930 still believed in the "vedic" ether. This means that his final conclusion must have been that all space is filled with this fluid ether and that submerged in this ether we find a second gaseous medium which we could refer to as "electricity".

It is important to know that Tesla believed in *universal* laws of nature, meaning that if we have a law that applies to gasses then it applies to all gasses *including electricity*.

True wireless

When people today hear the word "wireless", we immediately think of "radio controlled". But there are other means to create a wireless interaction. For example a T.V. remote control usually employs infra red light signals. Yet neither of these techniques would enable wireless power transmission over large distances. Tesla was fully aware of this and that is why he used an entirely different method. A method that he tried to explain many times, yet his attempts were in vain as people did not think of electricity in the same way as he did.

Having explained his view on electricity we can now understand his take on Earth resonance and his wireless system.

Let me first go to the pre-hearing interview of 1916 where Tesla had to explain his method to a lawyer, someone oblivious to advanced science.

I have an idea that [you] will get the best picture of the process in my system of transmission if you will imagine that the earth is a reservoir, say, of fluid under pressure -- that is the potential energy -- and at my plant, operating a distant tuned circuit, I must open a valve and enable that energy to flow in. It is exactly that way. The energy is all conserved, whether it is vibrating or purely potential. Whatever the transmitter does in the receiver, the effect is simply to open a valve, as it were, and permit energy to flow in.

Notice that there are two separate things; one, there is energy in the Earth and two, we need to open a valve to let this energy flow into the receiver. This is essential, for in the understanding of most people today the energy itself opens the valve as it were. Look for example at a radio receiver. The transmitter sends power at a certain frequency, the receiver - tuned to this frequency - automatically picks up this power.

Three years later in a magazine "Electrical Experimenter" Tesla goes into a bit more detail. But readers of this article especially today will still have that same handicap of thinking different about what electricity is.

February 1919

Famous Scientific Illusions

Part III. The Singular Misconception of the Wireless.

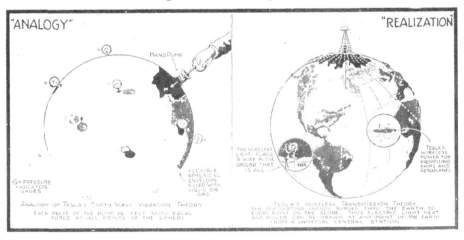

Imagine the earth to be a bag of rubber filled with water, a

small quantity of which is periodically forced in and out of the same by means of a reciprocating pump, as illustrated. If the strokes of the latter are effected in intervals of more than one hour and forty-eight minutes, sufficient for the transmission of the impulse through the whole mass, the entire bag will expand and contract and corresponding movements will be imparted to pressure gauges or movable pistons with the same intensity, irrespective of distance. By working the pump faster, shorter waves will be produced which, on reaching the opposite end of the bag, may be reflected and give rise to stationary nodes and loops, but in any case, the fluid being incompressible, its inclosure perfectly elastic, and the frequency of oscillations not very high, the energy will be economically transmitted and very little power consumed so long as no work is done in the receivers. This is a crude but correct representation of my wireless system in which, however, I resort to various refinements. Thus, for instance, the pump is made part of a resonant system of great inertia, enormously magnifying the force of the imprest impulses. The receiving devices are similarly conditioned and in this manner the amount of energy collected in them vastly increased.

The Earth now is not (just) filled with a fluid medium (the ether) but also with electricity; a gaseous medium. If we could rhythmically disturb that medium we could create a standing longitudinal wave in the Earth's electricity. Note that this is not just about moving large *charges* in and out of the Earth, it is about *electricity*, the cause of those charges.

If I make a huge Tesla coil and turn it on, it would pump electrical charges from the Earth in and out of its top-load, thus creating an alternating positive and negative high voltage.

But whether this voltage is positive or negative is irrelevant, as long as it is a high voltage it requires a lot of compressed electricity. Thus the impulse that we are sending into the Earth is not depending on the resonant frequency of our Tesla coil (as we found in the last set of key points concerning the diagram of the TMT), it depends on the rate at which we are switching the coil on and off, or - in a classical Tesla coil - on the spark-gap frequency. This is why we see so many patents for "circuit controllers" by Tesla; he wanted to get a precise control over this frequency. Today we could do that with transistors or vacuum tubes, but Tesla had to improvise.

This frequency has to match the Earth's frequency, the period of which can be calculated as follows:

>twice the Earth diameter divided by the speed of light

It is twice the diameter, because the impulse has to travel to the opposite end of the world and return. As the Earth is not a perfect globe the exact result depends on your location but is about 85 ms, corresponding to a frequency of about 11.8 Hz.

We can find these values in Tesla's own words in these articles.

1909, December 24th

Nikola Tesla's New Wireless

Mr. Tesla adds that in his experiments in Colorado it was shown that a very powerful current developed by the transmitter traversed the entire globe and returned to its origin in an interval of 84 one-thousandths of a second, this journey of 24,000 miles being effected almost without loss of energy.

1921, September 24th

Interplanetary Communication

While I am not prepared to discuss the various aspects of this subject at length, I may say that a skilful experimenter who is in the position to expend considerable money and time will undoubtedly detect waves of about 25,470,000 m.

To clarify the last quote: a wavelength of 25,470,000 m corresponds to a period of 84.9 ms or a frequency of 11.77 Hz.

I should be clear that this has nothing to do with Schumann-resonance nor with Zenneck-waves. The first is an effect that takes place in our atmosphere and the latter is a surface phenomenon, while the effect that Tesla refers to takes place *inside* the Earth and has been verified by my experiments in recent years.

On this large 11.8 Hz wave, that is gushing back and forth inside the Earth - this is the *energy* Tesla talked about in the first quote - one can modulate higher frequencies of positive and negative effects.

This large wave is electricity in its most rudimentary form, we are not talking about electrical charges. These higher frequency waves get amplified together with the low frequency wave thus creating a massive worldwide effect that can be tapped into at any point around the globe. The high frequency tone is the signal that 'opens the valve' in the receiver.

Some believe that Tesla planned to use the ionosphere to supply the return current. Though he did consider this plan, in the end he decided to only use the Earth as we can read in a birthday interview on July 10th, 1932.

Tesla Cosmic Ray Motor May Transmit Power 'Round Earth

He at that time announced two principles which could be used in this project. In one the ionizing of the upper air would make it as good a conductor of electricity as a metal. In the other the power would be transmitted by creating "standing waves" in the earth by charging the earth with a giant electrical oscillator that would make the earth vibrate electrically in the same way a bell vibrates mechanically when it is struck with a hammer.

"I do not use the plan involving the conductivity of the upper strata of the air," he said, "but I use the conductivity of the earth itself, and in this I need no wires to send electrical energy to any part of the globe."

Conclusions of part 1

What it a TMT?

The current consensus is that Tesla's patent 1,119,732 describes a Magnifying Transmitter. Tesla has been asked on several occasions to describe his Magnifying Transmitter and if it were true that this patent describes it, it would have been very easy for Tesla to just refer to that patent. Yet, he never does.

The patent simply describes a Tesla coil with a third so called "extra coil" added to it.

From Tesla's own words we can deduce beyond any reasonable doubt that this patent *does not* describe what Tesla would have called a Magnifying Transmitter. Besides that, we have actual physical evidence that these two are not the same.

From Tesla's writings and from the drawings that he left us we can establish with absolute certainty what the diagram of a Magnifying Transmitter would look like.

Excuse me putting so much stress on this fact, but since it does not align with the common consensus I believe this is very important conclusion.

What does a TMT do?

A Magnifying Transmitter transmits electrical power through the Earth, that can be picked up and converted into usable form by suitable receivers. This power can be modulated so that it conveys a message, thus it can also be used as a worldwide messaging system.

But this is what can be done with a machine as described in the before mentioned patent.

Also, why call it a *Magnifying* Transmitter, if it is just a transmitter?

There is reason to believe that this TMT does something more than just transmit power and/or messages, but Tesla does not talk about that in direct terms. We will explore this in part 2.

How does it do that?

Something is blocking us from understanding how the TMT works; and that is our understanding of what electricity is.

When *we* think of electricity we think of static or moving electrical charges.

When Tesla thinks of electricity he thinks of what *causes* these charges.

Tesla believes in a fluid homogeneous medium filling all space called the ether. Submerged in this ether he believes there must be a gaseous medium that he refers to as "the medium". It is this latter medium that Tesla thinks is causing electric and magnetic effects.

This "medium" clings to matter and therefore the Earth and its direct environment is filled with this "gas".

With the proper machinery you can create "sound waves" in this medium; waves of compressed "gas". This compressed "gas" represents energy and with suitable apparatus you can convert this energy into electric currents that can be used in households or industries.

In order for the receiver to be able to access this energy the TMT adds a high frequency note to the wave. This high frequency note "opens the valve" in the receiver and lets the energy flow in, as Tesla describes it in his own words.

In Tesla's "Art of Individualization" this note takes a more complex form and can be compared to a melody instead of a single tone. In this way receivers can be built that only respond to certain melodies and thus it becomes possible to send power or messages to specific receivers around the world.

Part 2: Diving Deeper

While every statement in part 1 was backed by a clear statement from Tesla himself we also need to entertain the thought that Tesla did not or could not tell us everything in clear and direct terms.

As I have said in the introduction, besides direct evidence we need to look at circumstantial evidence and finally we need to consider the possibility of a hidden message. I am 100% convinced that such a message exists, because I have found it, but to new-comers it may at first seem a little far fetched. Don't worry though, at some point the evidence can no longer be denied.

Please, dive along with me.

The Tesla code

Up until 1895 Tesla's written work is very precise and complete but after this date something changes. In March 1895 Tesla's laboratory in Houston street burned down and all of Tesla's work was lost. Whether it was this event or something else, it is very clear that Tesla's writing style changed after this date culminating into his masterpiece "the Problem of Increasing Human Energy", an article published in Century Illustrated Magazine in June 1900 with the help of his friend Robert Underwood Johnson.

This article appears to be a rather philosophical piece about many things outside of Tesla's field of expertise. Not at all what one would expect or hope to read from the world's foremost expert on electricity at his time. But after reading it and re-reading it several times a picture emerges; Tesla *is* writing about electricity but he does so using analogies. At one point he even gives you a key to understand the analogy. Let me show you these two quotes;

> *This new and inexhaustible source of food-supply will be of incalculable benefit to mankind, for it will enormously contribute to the increase of the human mass, and thus add immensely to human energy. Soon, I hope, the world will see the beginning of an industry which, in time to come, will, I*

> believe, be in importance next to that of iron.

Compare that with this quote from the same article:

> But for the time being, next to providing new resources of energy, it is of the greatest importance to making improvements in the manufacture and utilization of iron. Great advances are possible in these latter directions, which, if brought about, would enormously increase the useful performance of mankind.

It says exactly the same only an *"inexhaustible source of food-supply"* has been replaced with *"new resources of energy"*. From this I conclude that they stand for the same thing, the first being an analogy for the second. Also worth noting is the phase "next to the manufacture of iron", because that too is an analogy for the production of energy.

Recognizing this fact we now read the story of a triumphal man describing his discoveries and reasoning behind them.

It would go too far to completely dissect this article here as I have already done on steemit[11]. In this book I will limit myself to the most relevant and worthwhile facts.

Apart from this there are a number of patents dated after 1900 that show things outside of the field of electricity. I believe Tesla was trying to apply his electric discoveries to other areas. An example is patent 1,209,359: Speed Indicator.

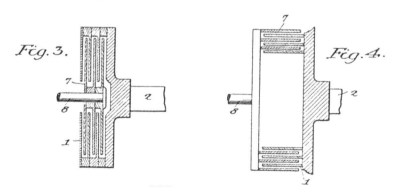

It is not difficult to recognize two different coil geometries Tesla used: the flat spiral and the cylindrical coil. As with his coils there is a

---

11 See https://steemit.com/science/@mage00000/reading-nikola-tesla-part-1-laying-a-foundation and subsequent articles.

primary initiating motion in a secondary.

In some cases however, he may have been trying to convey an idea. Let's turn to this one that I believe falls into this latter category.

Patent 1,113,716: Fountain

This is a patent that immediately stands out among all of Tesla's patents. It has very little to do with electricity nor anything practical or useful as *all* of his other patents do. Instead it shows how to create a fountain that efficiently pumps a lot of water around. Its purpose being a purely aesthetic one, yet Tesla still feels he has to put in some practical consideration as he adds:

> *The usual fountains are objectionable in many places on account of the facility they afford for the breeding of insects. The apparatus described not only makes this impossible but is a very efficient trap. Unlike the old devices in which only a very small volume of water is set in motion, such a waterfall is highly effective in cooling the surrounding atmosphere.*

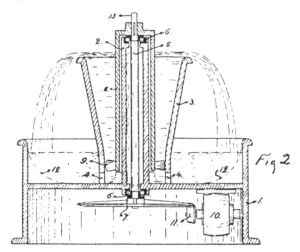

Well, yes, that may all be true but I have never seen a fountain using this principle in all of my years and all places that I travelled.

But what does it actually show? Below there is a reservoir of some fluid, in this case water. In this fluid, a rotation is established and because of this the fluid moves up.

If I use a tea spoon to stir my tea, I see the same effect; the tea level rises near the edges of my cup. Is this really the effect that Tesla wanted to claim in this patent. It almost feels silly.

However... Let's go back to where we were at the end of part 1 of this book. We had the Earth, being a huge reservoir of electricity in which we wanted to create some motion.

Now imagine this huge coil of the Colorado Springs experimental station, that you see in this picture behind Tesla. Through its wires electric currents are rushing in a circle.

Wouldn't that create a similar effect?

If you try to think like Tesla that the laws of physics are universal, then you can only come to one conclusion. It *must* do the same. The rotation created in the coil creates a vortex in electricity pulling up a large amount of it from the reservoir below. Exactly as described in the patent.

Although this does not in itself result in the high voltages seen in these pictures, it certainly facilitates it along with other effects.

Although my coils were not *this* big I believe my experimental results could certainly support this theory.

This concludes what can be said about wireless power transmission as proposed by Nikola Tesla and implemented in his Magnifying Transmitter. Now what can be said about the "Magnifying" bit?

You may have thought worldwide wireless power was interesting, well, this is where things get really fascinating!

A new source of energy

The question is "does the TMT provide a new source of energy?". To this question we do not get a direct answer but it is clear that Tesla was working on this during the same time that he was working on the development of the TMT, that is from 1889 to 1901.

1891, May 20th

> **Experiments with Alternate Currents of Very High Frequency and Their Application to Methods of Artificial Illumination**
>
> *For years the efforts of inventors have been directed towards obtaining electrical energy from heat by means of the thermopile. It might seem invidious to remark that but few know what is the real trouble with the thermopile. It is not the inefficiency or small output—though these are great drawbacks—but the fact that the thermopile has its phylloxera, that is, that by constant use it is deteriorated, which has thus far prevented its introduction on an industrial scale. Now that all modern research seems to point with certainty to the use of electricity of excessively high tension, the question must present itself to many whether it is not possible to obtain in a practicable manner this form of energy from heat. We have been used to look upon an electrostatic machine as a plaything, and somehow we couple with it the idea of the inefficient and impractical. But now we must think differently, for now we know that everywhere we have to deal with the same forces, and that it is a mere question of inventing proper methods or apparatus for rendering them available.*
>
> ---//---
>
> *But there is a possibility of obtaining energy not only in the form of light, but motive power, and energy of any other form, in some more direct way from the medium. The time will be when this will be accomplished, and the time has come when one may utter such words before an enlightened audience without being considered a visionary. We are whirling through endless space with an inconceivable speed, all around us everything is spinning, everything is moving, everywhere is*

energy. There must be some way of availing ourselves of this energy more directly. Then; with the light obtained from the medium, with the power derived from it, with every form of energy obtained without effort, from the store forever inexhaustible, humanity will advance with giant strides. The mere contemplation of these magnificent possibilities expands our minds, strengthen our hopes and fills our hearts with supreme delight.

Key points:
- the plan is to convert heat directly into high voltage electricity
- this heat has to come directly from "the medium"

In December 1931 Tesla refers to this statement in an interview

Our Future Motive Power

It was clear to me many years ago that a new and better source of power had to be discovered to meet the ever increasing demands of mankind. In a lecture delivered before the American Institute of Electrical Engineers at Columbia University May 20, 1891, I said: "We are whirling through endless space with inconceivable speed, all around us everything is spinning, everything is moving, everywhere is energy. There must be some way of availing ourselves of this energy more directly. Then, with the light obtained from the medium, with the power derived from it, with every form of energy obtained without effort, from the store forever inexhaustible, humanity will advance with giant strides." I have thought and worked with this object in view unremittingly and am glad to say that I have sufficient theoretical and experimental evidence to fill me with hope, not to say confidence, that my efforts of years will be rewarded and that we shall have at our disposal a new source of power, superior even to the hydro-electric, which may be obtained by means of simple apparatus everywhere and in almost constant and unlimited amount.

One may think that he refers to recent conclusions, but I don't think so mainly because he stopped experimenting a long time ago. I think his hopes are on someone to continue his work.

1892, February 3rd

Experiments with Alternate Currents of High Potential and High Frequency

Ere many generations pass, our machinery will be driven by a power obtainable at any point of the universe. This idea is not novel. Men have been led to it long ago by instinct or reason; it has been expressed in many ways, and in many places, in the history of old and new. We find it in the delightful myth of Antheus, who derives power from the earth; we find it among the subtle speculations of one of your splendid mathematicians and in many hints and statements of thinkers of the present time. Throughout space there is energy. Is this energy static or kinetic? If static our hopes are in vain; if kinetic—and this we know it is, for certain—then it is a mere question of time when men will succeed in attaching their machinery to the very wheelwork of nature. Of all, living or dead, Crookes came nearest to doing it. His radiometer will turn in the light of day and in the darkness of the night; it will turn everywhere where there is heat, and heat is everywhere. But, unfortunately, this beautiful little machine, while it goes down to posterity as the most interesting, must likewise be put on record as the most inefficient machine ever invented!

Key points:
- the source is "heat from the medium"
- Tesla mentions the work of Crookes, he does so more than once when discussing this subject

1893, February 24th

On Light and Other High Frequency Phenomena

So, the atom, the ulterior element of the Universe's structure, is tossed about in space eternally, a play to external influences, like a boat in a troubled sea. Were it to stop its motion it would die: matter at rest, if such a thing could exist, would be matter dead. Death of matter! Never has a sentence of deeper philosophical meaning been uttered. This is the way in which Prof. Dewar forcibly expresses it in the description of his admirable experiments, in which liquid oxygen is handled as one handles water, and air at ordinary pressure is made to condense and even to solidify by the intense cold: Experiments, which serve to illustrate, in his language, the last feeble manifestations of life, the last quiverings of matter about to die. But human eyes shall not witness such death. There is no death of matter, for throughout the infinite universe, all has to move, to vibrate, that is, to live.

Key points:
- matter can not "die" (cooled down to absolute zero). So you can extract endless heat energy from it?

An interesting piece is this one from March 8th, 1896. I will just copy a few highlights, because it is not written by Tesla himself.

Earth electricity to kill monopoly

The World is on the eve of an astounding revelation. The conditions under which we exist will be changed. The end has come to telegraph and telephone monopolies with a crash. Incidentally, all the other monopolies that depend on power of any kind will come to a sudden stop. The earth currents of electricity are to be harnessed. Nature supplies them free of charge.

---//---

Electricity will be as free as the air.

---//---

Monopolies for purveying steams power too will be forced to capitulate to free electricity, for with the latter manufactures will only have to connect their dynamos with the earth currents to set their machinery in motion. The successful adaptation of Tesla's discovery will administer a death-blow to the most galling slavery that has ever yoked the activities of

> *men to the treadmill of monopoly. Tesla is the wizard who is going to emancipate modern industries from the shackles of corrupting, dividend-grabbing, monopolistic corporations.*

You do not have to be a genius to understand that an idea like this will be violently opposed by these "monopolies". So much so that it will be entirely impossible to implement such a system. Though not written by Tesla, I do believe that it did come from Tesla only the one who wrote it did not understand much of what Tesla said.

"Where there is smoke, there is fire", they say...

1897, January 27th

On Electricity

> *We have to evolve means for obtaining energy from stores which are forever inexhaustible, to perfect methods which do not imply consumption and waste of any material whatever. Upon this great possibility, which I have long ago recognized, upon this great problem, the practical solution of which means so much for humanity, I have myself concentrated my efforts since a number of years, and a few happy ideas which came to me have inspired me to attempt the most difficult, and given me strength and courage in adversity.*

> *Nearly six years ago[12] my confidence had become strong enough to prompt me to an expression of hope in the ultimate solution of this all dominating problem. I have made progress since, and have passed the stage of mere conviction such as is derived from a diligent study of known facts, conclusions and calculations. I now feel sure that the realization of that idea is not far off. But precisely for this reason I feel impelled to point out here an important fact, which I hope will be remembered. Having examined for a long time the possibilities of the development I refer to, namely, that of the operation of engines on any point of the earth by the energy of the medium, I find that even under the theoretically best conditions such a method of obtaining power can not equal in economy, simplicity and many other features the present method, involving a conversion*

---

12 That must have been 1891

of the mechanical energy of running water into electrical energy and the transmission of the latter in the form of currents of very high tension to great distances. Provided, therefore, that we can avail ourselves of currents of sufficiently high tension, a waterfall affords us the most advantageous means of getting power from the sun sufficient for all our wants, and this recognition has impressed me strongly with the future importance of the water power, not so much because of its commercial value, though it may be very great, but chiefly because of its bearing upon our safety and welfare. I am glad to say that also in this latter direction my efforts have not been unsuccessful, for I have devised means which will allow us the use in power transmission of electromotive forces much higher than those practicable with ordinary apparatus. In fact, progress in this field has given me fresh hope that I shall see the fulfilment of one of my fondest dreams; namely, the transmission of power from station to station without the employment of any connecting wire. Still, whatever method of transmission be ultimately adopted, nearness to the source of power will remain an important advantage.

Tesla considers 2 options to solve the energy problem for humanity; one, a small machine to derive power from the environment, or two, a large machine that does so and transmits the energy through the natural media to small receivers. He goes back and forth between these two options a number of times.

As this is an excerpt from a speech held at Niagara Falls it is logical to focus on the transmission instead of local generation of power. We have already seen that in 1931 he states that he has a better solution.

These two options are described again in the next quote.

1898, November 30th

Tesla Describes his Efforts in Various Fields of Work

As to the idea of rendering the energy of the sun available for industrial purposes, it fascinated me early but I must admit it was only long after I discovered the rotating magnetic field that it took a firm hold upon my mind. In assailing the problem

> *I found two possible ways of solving it. Either power was to be developed on the spot by converting the energy of the sun's radiations or the energy of vast reservoirs was to be transmitted economically to any distance. Though there were other possible sources of economical power, only the two solutions mentioned offer the ideal feature of power being obtained without any consumption of material. After long thought I finally arrived at two solutions, but on the first of these, namely, that referring to the development of power in any locality from the sun's radiations, I can not dwell at present. The system of power transmission without wires, in the form in which I have described it recently, originated in this manner.*

Interesting statement! He *can not* dwell on the local development of power. Can not, or doesn't want to? In his article "the Problem of Increasing Human Energy" he mentions the same two options but then he does not want to talk about transmission.

From these quotes it becomes evident that he was thinking about providing a new and clean source of energy during this time so let's have a closer look at this time line.

The Tesla time line

Roughly up until 1888 Tesla was working on his induction motor, and AC power distribution. In that year Westinghouse hired Tesla for one year for the large fee of $2,000 ($52,900 in 2018's dollars) a month to be a consultant at the Westinghouse Electric & Manufacturing Company's Pittsburgh labs.

His patents and articles continue on this subject up until 1891. Then Tesla shows in his articles that he has been working on high frequency and high voltage currents. It appears Tesla's experiments focus on the question "what is electricity?". Is it a homogeneous fluid like the ether or is it a gaseous medium. We already know the outcome.

Tesla files a few patents on lighting systems and turns his attention to capacitor (called condenser in his time) and coil design and their application to create high frequency, high voltage currents.

Also he devotes a lot of time to enhance the spark-gap in a Tesla coil. He thinks too much energy is lost there and he wants to gain more control over its timing.

In March 1895 his laboratory burns down, destroying years of work and lots of valuable equipment.

After this Tesla's experiments and articles mainly focus on vacuum tubes and the production of Röntgen and Lenard rays.

In the second half of 1896 he patents the fruits of his labour, followed by a second wave around the end of 1897 and in first months of 1898.

At the Electrical Exhibition of 1898 in Madison Square Garden Tesla demonstrates his radio controlled boat and does so again in 1899 whilst travelling to Colorado Springs. There he could perform experiments with much higher voltages and produced the first man-made lightning. The thunder could be heard miles away.

In January 1900 Tesla returns and starts building his TMT at Wardenclyffe.

These are the years that I am interested in because here the TMT is developed. We put this time-line next to his autobiography and his most important article.

Tesla's autobiography

We turn to chapter 5, "The Magnifying Transmitter"[13]

As I review the events of my past life I realize how subtle are the influences that shape our destinies. An incident of my youth may serve to illustrate. One winter's day I managed to climb a steep mountain, in company with other boys. The snow was quite deep and a warm southerly wind made it just suitable for our purpose. We amused ourselves by throwing balls which would roll down a certain distance, gathering more or less snow, and we tried to outdo one another in this exciting sport. Suddenly a ball was seen to go beyond the limit, swelling to enormous proportions until it became as big as a house and plunged thundering into the valley below with a force that made the ground tremble. I looked on spellbound, incapable of understanding what had happened. For weeks afterwards the picture of the avalanche was before my eyes and I wondered how anything so small could grow to such an immense size. Ever since that time the magnification of feeble actions fascinated me, and when, years later, I took up the experimental study of mechanical and electrical resonance, I was keenly interested from the very start. Possibly, had it not been for that early powerful impression, I might not have followed up the little spark I obtained with my coil and never developed my best invention, the true history of which I'll tell here for the first time.

"Lionhunters" have often asked me which of my discoveries I prize most. This depends on the point of view. Not a few technical men, very able in their special departments, but dominated by a pedantic spirit and nearsighted, have asserted that excepting the induction motor I have given to the world

---
13   Published in the June issue of the Electrical Experimenter in 1919

little of practical use. This is a grievous mistake. A new idea must not be judged by its immediate results.

My alternating system of power transmission came at a psychological moment, as a long-sought answer to pressing industrial questions, and although considerable resistance had to be overcome and opposing interests reconciled, as usual, the commercial introduction could not be long delayed.

Now, compare this situation with that confronting my turbine, for example. One should think that so simple and beautiful an invention, possessing many features of an ideal motor, should be adopted at once and, undoubtedly, it would under similar conditions. But the prospective effect of the rotating field was not to render worthless existing machinery; on the contrary, it was to give it additional value. The system lent itself to new enterprise as well as to improvement of the old. My turbine is an advance of a character entirely different. It is a radical departure in the sense that its success would mean the abandonment of the antiquated types of prime movers on which billions of dollars have been spent. Under such circumstances the progress must needs be slow and perhaps the greatest impediment is encountered in the prejudicial opinions created in the minds of experts by organized opposition.

Only the other day I had a disheartening experience when I met my friend and former assistant, Charles F. Scott, now professor of Electrical Engineering at Yale. I had not seen him for a long time and was glad to have an opportunity for a little chat at my office. Our conversation naturally enough drifted on my turbine and I became heated to a high degree. "Scott," I exclaimed, carried away by the vision of a glorious future, "my turbine will scrap all the heat-engines in the world." Scott stroked his chin and

looked away thoughtfully, as though making a mental calculation. "That will make quite a pile of scrap," he said, and left without another word!

These and other inventions of mine, however, were nothing more than steps forward in certain directions. In evolving them I simply followed the inborn sense to improve the present devices without any special thought of our far more imperative necessities. The "Magnifying Transmitter" was the product of labours extending through years, having for their chief object the solution of problems which are infinitely more important to mankind than mere industrial development.

If my memory serves me right, it was in November, 1890, that I performed a laboratory experiment which was one of the most extraordinary and spectacular ever recorded in the annals of science. In investigating the behaviour of high frequency currents I had satisfied myself that an electric field of sufficient intensity could be produced in a room to light up electrodeless vacuum tubes. Accordingly, a transformer was built to test the theory and the first trial proved a marvellous success. It is difficult to appreciate what those strange phenomena meant at that time. We crave for new sensations but soon become indifferent to them. The wonders of yesterday are today common occurrences. When my tubes were first publicly exhibited they were viewed with amazement impossible to describe. From all parts of the world I received urgent invitations and numerous honours and other flattering inducements were offered to me, which I declined.

But in 1892 the demands became irresistible and I went to London where I delivered a lecture before the Institution of Electrical Engineers. It had been my intention to leave immediately for Paris in compliance with a similar obligation, but Sir James Dewar insisted on my appearing before the Royal Institution. I was a man of firm resolve but succumbed easily to the forceful arguments of the great Scotsman. He pushed me into a chair and poured out half a glass of a wonderful brown fluid which sparkled in all sorts of iridescent colours and tasted like nectar. "Now," said he. "you are sitting in Faraday's chair and you are enjoying whiskey he used to drink." In both aspects it was an enviable experience. The next evening I gave a demonstration before that Institution, at the termination of which Lord Rayleigh addressed the audience and his generous words gave me the first start in these endeavours. I fled from London and later from Paris to escape favours showered upon me, and journeyed to my home where I passed through a most painful ordeal and illness. Upon regaining my health I began to formulate plans for the resumption of work in America. Up to that time I never realized that I possessed any particular gift of discovery but Lord Rayleigh, whom I always considered as an ideal man of science, had said so and if that was the case I felt that I should concentrate on some big idea.

One day, as I was roaming in the mountains, I sought shelter from an approaching storm. The sky became overhung with heavy clouds but somehow the rain was delayed until, all of a sudden, there was a lightning flash and a few moments after a deluge. This observation set me thinking. It was manifest that the two phenomena were closely related, as cause and effect, and a little reflection led me to the conclusion that the electrical energy involved in the precipitation of the water was inconsiderable, the function of lightning being much like that of a sensitive trigger.

Here was a stupendous possibility of achievement. If we could produce electric effects of the required quality, this whole planet and the conditions of existence on it could be transformed. The sun raises the water of the oceans and winds drive it to distant regions where it remains in a state of most delicate balance. If it were in our power to upset it when and wherever desired, this mighty life-sustaining stream could be at will controlled. We could irrigate arid deserts, create lakes and rivers and provide motive power in unlimited amounts. This would be the most efficient way of harnessing the sun to the uses of man. The consummation depended on our ability to develop electric forces of the order of those in nature. It seemed a hopeless undertaking, but I made up my mind to try it and immediately on my return to the United States, in the Summer of 1892, work was begun which was to me all the more attractive, because a means of the same kind was necessary for the successful transmission of energy without wires.

The first gratifying result was obtained in the spring of the succeeding year when I reached tensions of about 1,000,000 volts with my conical coil. That was not much in the light of the present art, but it was then considered a feat. Steady progress was made until the destruction of my laboratory by fire in 1895, as may be judged from an article by T. C. Martin which appeared in the April number of the Century Magazine. This calamity set me back in many ways and most of that year had to be devoted to planning and reconstruction. However, as soon as circumstances permitted, I returned to the task.

Although I knew that higher electro-motive forces were attainable with apparatus of larger dimensions, I had an instinctive perception that the object could be accomplished by the proper design of a comparatively small and compact transformer. In carrying on tests with a secondary in the form of a flat spiral, as illustrated in my patents, the absence of streamers surprised me, and it was not long before I discovered that this was due to the position of the turns and their mutual action. Profiting from this observation I resorted to the use of a high tension conductor with turns of considerable diameter sufficiently separated to keep down the distributed capacity, while at the same time preventing undue accumulation of the charge at any point. The application of this principle enabled me to produce pressures of 4,000,000 volts, which was about the limit obtainable in my new laboratory at Houston Street, as the discharges extended through a distance of 16 feet. A photograph of this transmitter was published in the Electrical Review of November, 1898.[14]

In order to advance further along this line I had to go into the open, and in the spring of 1899, having completed preparations for the erection of a wireless plant, I went to Colorado where I remained for more than one year. Here I introduced other improvements and refinements which made it

---

[14] I could find two such articles, one of the 16th and one of the 30th, both copies did not include a picture. I believe Tesla refers to his flat spiral coil here.

possible to generate currents of any tension that may be desired. Those who are interested will find some information in regard to the experiments I conducted there in my article, "The Problem of Increasing Human Energy" in the Century Magazine of June, 1900, to which I have referred on a previous occasion.

Note that this chapter starts with a story about creating an avalanche and how this inspired him to develop a little spark into his TMT. Tesla continues with these points:

- the importance of the TMT
- his discovery of electrode-less vacuum tubes
- he should concentrate on some big idea
- lightning as a trigger
- same requirements as wireless power, so not just wireless power
- 1 MV reached in spring 1893 with conical coil
- tests with a flat spiral coil after 1895 producing 4 MV
- his final experiments in Colorado Springs

This quote ends with a reference to his article "The Problem of Increasing Human Energy". Tesla refers to this article more than to any other of his articles. Let's look at these two chapters thereof.

Increasing Human Energy

A DEPARTURE FROM KNOWN METHODS—POSSIBILITY OF A "SELF-ACTING" ENGINE OR MACHINE, INANIMATE, YET CAPABLE, LIKE A LIVING BEING, OF DERIVING ENERGY FROM THE MEDIUM—THE IDEAL WAY OF OBTAINING MOTIVE POWER.

When I began the investigation of the subject under consideration, and when the preceding or similar ideas presented themselves to me for the first time, though I was then unacquainted with a number of the facts mentioned, a survey of the various ways of utilizing the energy of the medium convinced me, nevertheless, that to arrive at a thoroughly satisfactory practical solution a radical departure from the methods then known had to be made. The windmill, the solar engine, the engine driven by terrestrial heat, had their limitations in the amount of power obtainable. Some new way had to be discovered which would enable us to get more energy. There was enough heat-energy in the medium, but only a small part of it was available for the operation of an engine in the ways then known. Besides, the energy was obtainable only at a very slow rate. Clearly, then, the problem was to discover some new method which would make it possible both to utilize more of the heat-energy of the medium and also to draw it away from the same at a more rapid rate.

I was vainly endeavouring to form an idea of how this might be accomplished, when I read some statements from Carnot and Lord Kelvin (then Sir William Thomson) which meant virtually that it is impossible for an inanimate mechanism or self-acting machine to cool a portion of the medium below the temperature of the surrounding, and operate by the heat abstracted. These statements interested me intensely. Evidently a living being could do this very thing, and since the experiences of my early life which I have related had convinced me that a living being is only an automaton, or, otherwise stated, a "self-acting-engine," I came to the conclusion that it was possible to construct a machine which would do the same. As the first step toward this realization I conceived the following mechanism. Imagine a thermopile consisting of a number of bars of metal extending from the earth to the outer space beyond the atmosphere. The heat from below, conducted upward along these metal bars, would cool the earth or the sea or the air, according to the location of the lower parts of the bars, and the result, as is well known, would be an electric current circulating in these bars. The two terminals of the thermopile could now be joined through an electric motor, and, theoretically, this motor would run on and on, until the media below would be cooled down to the temperature of the outer space. This would be an inanimate engine which, to all evidence, would be cooling a portion of the medium below the temperature of the surrounding, and operating by the heat abstracted.

DIAGRAM b.
OBTAINING ENERGY FROM THE AMBIENT MEDIUM
A, medium with little energy; B, B, ambient medium with much energy; O, path of the energy.

But was it not possible to realize a similar condition without necessarily going to a height? Conceive, for the sake of illustration, [a cylindrical] enclosure T, as illustrated in diagram b, such that energy could not be transferred across it except through a channel or path O, and that, by some means or other, in this enclosure a medium were maintained which would have little energy, and that on the outer side of the same there would be the ordinary ambient medium with much energy. Under these assumptions the energy would flow through the path O, as indicated by the arrow, and might then be converted on its passage into some other form of energy. The question was, Could such a condition be attained? Could we produce artificially such a "sink" for the energy of the ambient medium to flow in? Suppose that an extremely low temperature could be maintained by some process in a given space; the surrounding medium would then be compelled to give off heat, which could be converted into mechanical or other form of energy, and utilized. By realizing such a plan, we should be enabled to get at any point of the globe a continuous supply of energy, day and night. More than this, reasoning in the abstract, it would seem possible to cause a quick circulation of the medium, and thus draw the energy at a very rapid rate.

Here, then, was an idea which, if realizable, afforded a happy solution of the problem of getting energy from the medium. But was it realizable? I convinced myself that it was so in a number of ways, of which one is the following. As regards heat, we are at a high level, which may be represented by the surface of a mountain lake considerably above the sea, the level of which may mark the absolute zero of temperature existing in the interstellar space. Heat, like water, flows from high to low level, and, consequently, just as we can let the water of the lake run down to the sea, so we are able to let heat from the earth's surface travel up into the cold region above. Heat, like water, can perform work in flowing down, and if we had any doubt as to whether we could derive energy from the medium by means of a thermopile, as before described, it would be dispelled by this analogue. But can we produce cold in a given portion of the space and cause the heat to flow in continually? To create such a "sink," or "cold hole," as we might say, in the medium, would be equivalent to producing in the lake a space either empty or filled with something much lighter than water. This we could do by placing in the lake a tank, and pumping all the water out of the latter. We know, then, that the water, if allowed to flow back into the tank, would, theoretically, be able to perform exactly the same amount of work which was used in pumping it out, but not a bit more. Consequently nothing could be gained in this double operation of first raising the water and then letting it fall down. This would mean that it is impossible to create such a sink in the medium. But let us reflect a moment. Heat, though following certain general laws of mechanics, like a fluid, is not such; it is energy which may be converted into other forms of energy as it passes from a high to a low level. To make our mechanical analogy complete and true, we must, therefore, assume that the water, in its passage into the tank, is converted into something else, which may be taken out of it without using any, or by using very little, power. For example, if heat be represented in this analogue by the water of the lake, the oxygen and hydrogen composing the water may illustrate other forms of energy into which the heat is transformed in passing from hot to cold. If the process of heat transformation were absolutely perfect, no heat at all would arrive at the low level, since all of it would be converted into other forms of energy. Corresponding to this ideal case, all the water flowing into the tank would be decomposed into oxygen and hydrogen before reaching the bottom, and the result would be that water would continually flow in, and yet the tank would remain entirely empty, the gases formed escaping. We would thus produce, by expending initially a certain amount of work to create a sink for the heat or, respectively, the water to flow in, a condition enabling us to get any amount of energy without further effort. This would be an ideal way of obtaining motive power. We do not know of any such absolutely perfect process of heat-conversion, and consequently some heat will generally reach the low level, which means to say, in our mechanical analogue, that some water will arrive at the bottom of the tank, and a gradual and slow filling of the latter will take place, necessitating continuous pumping out. But evidently there will be less to pump out than flows in, or, in other words, less energy will be needed to maintain the initial condition than is developed by the fall, and this is to say that some energy will be gained from the medium. What is not converted in flowing down can just be raised up with its own energy, and what is converted is clear gain. Thus the virtue of the principle I have discovered resides wholly in the conversion of the energy on the downward flow.

FIRST EFFORTS TO PRODUCE THE SELF-ACTING ENGINE—THE MECHANICAL OSCILLATOR—WORK OF DEWAR AND LINDE—LIQUID AIR.

Having recognized this truth, I began to devise means for carrying out my idea, and, after long thought, I finally conceived a combination of apparatus which should make possible the obtaining of power from the medium by a process of continuous cooling of atmospheric air. This apparatus, by continually transforming heat into mechanical work, tended to become colder and colder, and if it only were practicable to reach a very low temperature in this manner, then a sink for the heat could be produced, and energy could be derived from the medium. This seemed to be contrary to the statements of Carnot and Lord Kelvin before referred to, but I concluded from the theory of the process that such a result could be attained. This conclusion I reached, I think, in the latter part of 1883, when I was in Paris, and it was at a time when my mind was being more and more dominated by an invention which I had evolved during the preceding year, and which has since become known under the name of the "rotating magnetic field." During the few years which followed I elaborated further the plan I had imagined, and studied the working conditions, but made little headway. The commercial introduction in this country of the invention before referred to required most of my energies until 1889, when I again took up the idea of the self-acting machine. A closer investigation of the principles involved, and calculation, now showed that the result I aimed at could not be reached in a practical manner by ordinary machinery, as I had in the beginning expected. This led me, as a next step, to the study of a type of engine generally designated as "turbine," which at first seemed to offer better chances for a realization of the idea. Soon I found, however, that the turbine, too, was unsuitable. But my conclusions showed that if an engine of a peculiar kind could be brought to a high degree of perfection, the plan I had conceived was realizable, and I resolved to proceed with the development of such an engine, the primary object of which was to secure the greatest economy of transformation of heat into mechanical energy. A characteristic feature of the engine was that the work-performing piston was not connected with anything else, but was perfectly free to vibrate at an enormous rate. The mechanical difficulties encountered in the construction of this engine were greater than I had anticipated, and I made slow progress. This work was continued until early in 1892, when I went to London, where I saw Professor Dewar's admirable experiments with liquefied gases. Others had liquefied gases before, and notably Ozlewski[15] and Pictet had performed creditable early experiments in this line, but there was such a vigour about the work of Dewar that even the old appeared new. His experiments showed, though in a way different from that I had imagined, that it was possible to reach a very low temperature by transforming heat into mechanical work, and I returned, deeply impressed with what I had seen, and more than ever convinced that my plan was practicable. The work temporarily interrupted was taken up anew, and soon I had in a fair state of perfection the engine which I have named "the mechanical oscillator." In this machine I succeeded in doing away with all packings, valves, and lubrication, and in producing so rapid a vibration of the piston that shafts of tough steel, fastened to the same and vibrated longitudinally, were torn asunder. By combining this engine with a dynamo of special design I produced a highly efficient electrical generator, invaluable in measurements and determinations of

---

15   Karol Stanisław Olszewski (1846-1915)

physical quantities on account of the unvarying rate of oscillation obtainable by its means. I exhibited several types of this machine, named "mechanical and electrical oscillator," before the Electrical Congress at the World's Fair in Chicago during the summer of 1893, in a lecture which, on account of other pressing work, I was unable to prepare for publication. On that occasion I exposed the principles of the mechanical oscillator, but the original purpose of this machine is explained here for the first time.

In the process, as I had primarily conceived it, for the utilization of the energy of the ambient medium, there were five essential elements in combination, and each of these had to be newly designed and perfected, as no such machines existed. The mechanical oscillator was the first element of this combination, and having perfected this, I turned to the next, which was an air-compressor of a design in certain respects resembling that of the mechanical oscillator. Similar difficulties in the construction were again encountered, but the work was pushed vigourously, and at the close of 1894 I had completed these two elements of the combination, and thus produced an apparatus for compressing air, virtually to any desired pressure, incomparably simpler, smaller, and more efficient than the ordinary. I was just beginning work on the third element, which together with the first two would give a refrigerating machine of exceptional efficiency and simplicity, when a misfortune befell me in the burning of my laboratory, which crippled my labours and delayed me. Shortly afterwards Dr. Carl Linde announced the liquefaction of air by a self-cooling process, demonstrating that it was practicable to proceed with the cooling until liquefaction of the air took place. This was the only experimental proof which I was still wanting that energy was obtainable from the medium in the manner contemplated by me.

Key points:
- Self-acting engine deriving heat from the medium
- conversion of the energy on the downward flow
- tests with a turbine

- the primary object of which was to secure the greatest economy of transformation of heat into mechanical energy. A characteristic feature of the engine was that the work-performing piston was not connected with anything else, but was perfectly free to vibrate at an enormous rate
- Professor Dewar's experiments with liquefied gases
- there were five essential elements in combination
- mechanical oscillator (1893)
- an air-compressor of a design in certain respects resembling that of the mechanical oscillator (end of 1894)
- I was just beginning work on the third element, which together with the first two would give a refrigerating machine of exceptional efficiency and simplicity, when a misfortune befell me in the burning of my laboratory (1895)
- Dr. Carl Linde announced the liquefaction of air by a self-cooling process (regenerative cooling)

With these quotes one has to bear in mind that Tesla is writing in analogies here. To fully understand them we must find what they stand for.

Combining the pieces

In his article we recognize Tesla's very systematic approach and for that reason I believe it is unlikely that Tesla was working on more than one large project at the same time. He said he was working on a "self-acting engine" (analogy) consisting of 5 parts. He also was working on his TMT during this time, so the idea suggests itself that the TMT was a "self-acting engine".

Let us look at a diagram of Linde's cooling process.

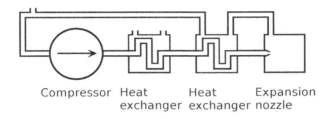

Compressor   Heat   Heat   Expansion
             exchanger   exchanger   nozzle

It consists of a compressor with heat exchanger, then a second heat exchanger for regenerative cooling and next the expansion nozzle.

An ordinary cooling system, such as in a fridge or airco does not have the second heat exchanger. It operates on a gaseous medium that gets compressed. Because of this compression the internal heat of the gas is concentrated making the gas hot. This heat is exchanged with the environment, after that the gas has less internal heat. Therefore, when it expands to its original pressure, it cools.

If a gas can do this, then electricity -being a gaseous medium- can do the same. A compressor consists of a motor (oscillator) and a pump, thus we have 3 elements:

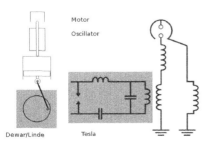

The motor creates a reciprocating movement. It can be seen as an oscillator.

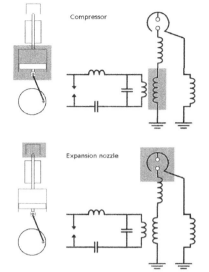

Next we have a pump that compresses the working medium, gas in Dewar's case or electricity in Tesla's.

To make this into a refrigerating machine the next thing we need is an expansion nozzle.

Tesla started working on this around 1895 when we notice his interest in vacuum tubes, x-rays and Lenard rays.

This is not a very straight-forward thing as this vacuum tube had to be able to withstand many millions of volts. Tesla mentions that he struggled with this until he found the ultimate and perfect solution suitable for any voltage no matter how high. That sounds quite amazing, if true.

Since this is a complex issue let's take a step back here and look where this idea could have originated.

Sir William Crookes

Tesla mentions his name quite often and most notably here on February 3rd, 1892

Experiments with Alternate Currents of High Potential and High Frequency

I need not mention many names which are world-known—names of those among you who are recognized as the leaders in this enchanting science; but one, at least, I must mention—a name which could not be omitted in a demonstration of this kind. It is a name associated with the most beautiful invention ever made: it is Crookes!

When I was at college, a good time ago; I read, in a translation (for then I was not familiar with your magnificent language), the description of his experiments on radiant matter. I read it only once in my life—that time—yet every detail about that charming work I can remember this day. Few are the books, let me say, which can make such an impression upon the mind of a student.

Crookes in turn was inspired by Faraday who believed there should be a 4th state of matter (beyond solid, liquid and gaseous) which he called "radiant". Crookes reasoned that phases of matter differ mostly in density, and so by reducing the pressure of a gas we may find this 4th state. He started experimenting with highly evacuated tubes, passing electric currents through them and found some truly remarkable facts. He concluded that the negative electrode (cathode) would emit rays perpendicular to its surface. (later he found that also the anode emitted some sort of rays) He assumed that these rays were the radiant state of the residual gasses inside the tube.

Something very remarkable happens here that could not have escaped Tesla: these rays are created by a potential difference between the anode (positive terminal) and cathode (negative terminal), yet these rays do not necessarily go to the anode. This effect matches the analogy that we just read in "the Problem of Increasing Human Energy" with the "conversion of the downward flow".

Namely, the electrical energy from the cathode is attracted by the anode but on its way there it gets converted into a "radiant state" and never arrives at the anode. Thus a positive charge on the anode would never stop to produce this radiant matter.

This radiant matter can be converted into electrical energy and this completes the analogy. Tesla experiments with this concept for many years but never gets more than a "feeble current" (as described in his patents on this work: 685,957 and 685,958 Apparatus for and Method of Utilizing Radiant Energy). Also operating a vacuum tube on voltages well over a million leads to insurmountable problems.

A new kind of vacuum tube needed to be designed.

The open vacuum tube

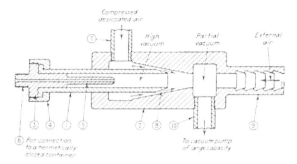

In an article of May 16th, 1935 "The New Art of Projecting Concentrated Non-dispersive Energy Through Natural Media" Tesla describes his "open vacuum tube".

Although I agree that in this way a low pressure can be attained, I do not think that you will obtain a high vacuum such as needed for Crookes experiments. But we need to remember the "Tesla Code". Tesla could be describing an electrical phenomenon using this analogy. Let's just consider this possibility and see where it leads.

Electricity, especially in the case of high frequencies, flows along a conductor and experiences drag (resistance) inside of it. This is demonstrated by the "skin-effect" and the fact that both electric and magnetic effects take place outside of the conductor. So what would an electric discharge from a metallic conductor look like?

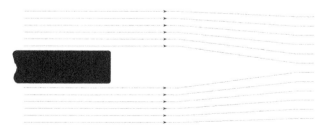

This would create the exact same effect; a low pressure region at the beginning of the discharge. Given the enormous velocity of electricity this low pressure region may indeed get close to a vacuum.

If this is true, any electrical discharge could be seen as a vacuum tube. A vacuum tube of extreme simplicity and suitable for any voltage exactly as Tesla says.

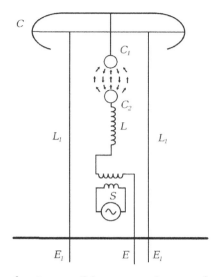

Returning to our original diagram, the expansion nozzle would then become a simple electric discharge and would be located between C_1 and C_2.

A high frequency would help increase the vacuum and a high voltage (pressure) would also increase the cooling effect.

Tesla specifically mentions this in his article[16] when talking about "burning atmospheric nitrogen".

This of course is an analogy. This "nitrogen" is electricity that needs to be "burned" (polarized) to make it possible to use it as a fertilizer to fertilize the Earth so that (power-) plants can grow, which in turn provide food (energy) for humanity. This is what the TMT does, it extracts electricity from the air and uses the Earth to distribute it to the receiving plants world-wide, which in turn transform it into a usable form.

Tesla jokingly adds:

> *Without wishing to put myself on record as a prophet, I do not hesitate to say that the next years will see the establishment of an "air-power," and its centre may be not far from New York.*

Of course referring to his project at Wardenclyffe, not to some air force.

The "extra coil"

Before turning to element 4 and 5, I must add an extra note about the extra coil.

To achieve very high potentials Tesla employs a Tesla coil with 3 coils; a primary, a secondary and a so called "extra coil". The primary coil is part of an oscillator circuit and creates a very powerful alternating magnetic field. This field induces an electro-motive force in each of the windings of the secondary coil. These add up to a very

---

16 "The Problem of Increasing Human Energy"

high alternating electric potential at the top of the secondary.

Tesla noted that this energy transfer happens both ways, from the primary to the secondary and back again. The latter is undesirable if one wants to obtain a great resonant rise. This led to experimenting with different coil geometries and ultimately to splitting the secondary coil into two coils. The first receives energy from the primary, and in the second a great resonant rise is established.

This resonant rise can be seen as pushing someone sitting on a swing. Each well timed push adds a little energy making the swing go higher. From this it is easy to see that this delays the rise as well as increases it. This creates a distinct difference between a 2-coil system and a 3-coil system.

A 2-coil system immediately gives a high voltage when the spark-gap fires and this voltage gradually diminishes until the spark-gap fires again.

In a 3-coil system the voltage gradually increases after the spark-gap fires until there is not enough energy left in the first two coils to increase the energy in the third. From that point on the voltage would decrease again, but if we can replenish this energy by having the spark-gap fire again, then we can establish a slow continuous rise in the "extra coil".

This is an important thing to realize, because without it we would not have created a "refrigerating machine".

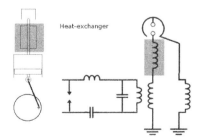

The relatively slow rise in voltage (read: electrical pressure) gives time for the developed heat to be dissipated to the environment. Thus the "extra coil" becomes our first heat-exchanger.

Then after a slow rise, we get a discharge (in our expansion nozzle) resulting in a sudden drop of pressure and thus the desired cooling effect. This exact same pattern is known to create very strange and long sword-like sparks with vacuum tube Tesla coils.

Elements 4 and 5

Tesla gives us the first 3 elements but of the last 2 he only says that the experiments of Linde were the only experimental proof which he was still wanting. The only thing Linde added was a regenerative loop, so that is what these final 2 elements should add. We also know from the diagram that we already have, how these two elements are implemented.

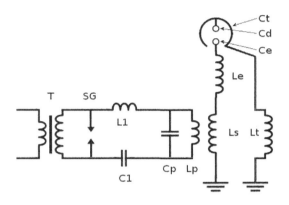

They are C_t and L_t, a system that Tesla refers to as the "free system" and which is most likely what Tesla called "the work-performing piston, not connected with anything else, but perfectly free to vibrate at an enormous rate". This is most likely the fourth element.

If this system is tuned to the same frequency as the primary system (C_p and L_p) then we can create a regenerative loop by adding the Earth as the 5th and final element.

What happens is this:

- the primary system initiates an oscillation in the secondary system (L_s, L_e and C_e)
- this results in an increasing voltage at C_e, which will eventually discharge to C_d.
- this makes the "free system" oscillate at its resonant frequency
- this sends an impulse into the Earth which will return about 85 ms later, …

- … increasing the energy in our secondary system

Now, if energy is somehow added to this system we know where that would happen (between C_e and C_d), but how? And where would that energy come from?

Cosmic rays

2ⁿᵈ Law of Thermodynamics

A self-acting-engine would seem to go against our fundamental laws of physics. How did Tesla see this? For this we turn to an article from October 13th, 1932.

> **The Eternal Source of Energy of the Universe, Origin and Intensity of Cosmic Rays**
>
> *In a communication to the Royal Society of Edinburgh dated April 19, 1852 and the Philosophical Magazine of October of the same year, Lord Kelvin drew attention to the general tendency in nature towards dissipation of mechanical energy, a fact borne out in daily observation of thermo-dynamic and dynamo-thermic processes and one of ominous significance. It meant that the driving force of the universe was steadily decreasing and that ultimately all of its motive energy will be exhausted none remaining available for mechanical work. In the macro-cosmos, with its countless conception, this process might require billion of years for its consummation; but in the infinitesimal worlds of the micro-cosmos it must have been quickly completed. Such being the case then, according to an experimental findings and deductions of positive science, any material substance (cooled down to the absolute zero of temperature) should be devoid of an internal movement and energy, so to speak, dead.*
>
> *This idea of the great philosopher, who later honoured me with his friendship, had a fascinating effect on my mind and in meditating over it I was struck by the thought that if there is energy within the substance it can only come from without. This truth was so manifest to me that I expressed it in the following axiom: "There is no energy in matter except that absorbed from the medium." Lord Kelvin gave us a picture of a*

dying universe, of a clockwork wound up and running down, inevitably doomed to come to a full stop in the far, far off future. It was a gloomy view incompatible with artistic, scientific and mechanical sense. I asked myself again and again, was there not some force winding up the clock as it runs down? The axiom I had formulated gave me a clue. If all energy is supplied to matter from without then this all important function must be performed by the medium. Yes--but how?

I pondered over this oldest and greatest of all riddles of physical science a long time in vain, despairingly remind of the words of the poet:

> *Wo fass ich dich unendliche Natür?*
> *Euch Brüste wo Ihr Quellen alles Lebens*
> *An denen Himmel und Erdë hangt...*

> *Where, boundless nature, can I hold you fast?*
> *And where you breasts? Wells that sustain*
> *All life--the heaven and the earth are nursed.*
> *Goethe. Faust*

What I strove for seemed unattainable, but a kind fate favored me and a few inspired experiments lifted the veil. It was a revelation wonderful and incredible explaining many mysteries of nature and disclosing as in a lightening flash the illusionary character of some modem theories incidentally also bearing out the universal truth of the above axiom.

We read that Tesla knew of this law, but could not believe that this was the complete truth. There had to be "some force to wind the clock up again".

He concluded that electrodes when brought to a sufficiently high potential in a vacuum would emit particle rays which he called primary cosmic rays (I shall refer to these as PCR). This he was able to verify in his laboratory as he says in this quote from the same article.

> *Rays in every respect similar to the cosmic are produced by my vacuum tubes when operated at pressures of ten millions of*

> *volts or more, but even if it were not confirmed by experiment, the theory I advanced in 1897 would afford the simplest and most probable explanation of the phenomena. Is not the universe with its infinite and impenetrable boundary a perfect vacuum tube of dimensions and power inconceivable? Are not its fiery suns electrodes at temperatures far beyond any we can apply in the puny and crude contrivances of our making? Is it not a fact that the suns and stars are under immense electrical pressures[17] transcending any that man can ever produce and is this not equally true of the vacuum in celestial space? Finally, can there be any doubt that cosmic dust and meteoric matter present an infinitude of targets acting as reflectors and transformers of energy? If under ideal working conditions, and with apparatus on a scale beyond the grasp of the human mind, rays of surpassing intensity and penetrative power would not be generated, then, indeed, nature has made an unique exception to its laws.*

These rays are the reason for radioactive decay and for secondary cosmic rays (SCR) when they interact with matter[18].

Modern science does not recognize Tesla's PCR, but *does* recognize SCR which we call Cosmic Background Radiation (CBR) and are assumed to be the remnants of the Big Bang.

These PCR are the force that wind up the clock as it runs down and thus go against our 2nd Law of Thermodynamics.

---

17 Tesla had established that our Sun sits at an electric potential of 216 billion volt as he says in an article of August 18th, 1935 and various articles after that date
18 Think of bremsstrahlung

Energy from cosmic rays

To get an idea of the available energy in PCR Tesla says on August 18th, *1935:*

Expanding Sun Will Explode Some Day Tesla Predicts

The greatest mistake is made in the appraisal of the energy of cosmic rays. In most cases the ionizing action is used as a criterion, which is useless, for the most powerful cosmic rays virtually do not ionize at all and leave no trace of their passage through the instrument. I have resorted to different means and methods and have found that the energy of the cosmic radiations impinging upon the earth from all sides is stupendous, such that if all of it were converted into heat the globe quickly would be melted and volatilized.

But how to access this energy?

Tesla explains on his birthday in 1932.

Tesla Cosmic Ray Motor May Transmit Power 'Round Earth

"I will tell you in the most general way," he said. *"The cosmic ray ionizes the air, setting free many charges—ions and electrons. These charges are captured in a condenser which is made to discharge through the circuit of the motor."*

As this is after 1895 we have to consider the fact that he uses analogies. It would appear to me that this "motor" is the "free system" of the TMT. This "motor" creates a motion in the Earth's electricity and thus transmits power 'round the world.

So this is how it is done. The TMT cools a portion of the medium inside an electric discharge. This cooled medium absorbs heat energy from the PCR, probably by creating "whirls" which manifest as electric charges[19]. These new charges are caught in the flow of charges in the discharge and thus end up on the receiving electrode.

Thus more charges and more energy arrives at this receiving electrode than left the sending electrode. This extra energy was supplied by the PCR.

---

19   I'll elaborate on this in the appendices as this is speculation from my part.

One story told 5 times

There is one story that Tesla tells on 5 different occasions, so it must be important. Let's take a look.

1904, June 17th

Patent Application 213,055[20]

It is well known that through the sun's heat water is evaporated and elevated to more or less considerable heights, where it remains in a state of delicate suspension, until disturbing influences arise which cause a condensation of vapour particles and their fall to the earth's surface under the action of gravity. Thus all vegetation and animal life on this globe is sustained. The body of water maintained in the air is like a heavy weight in labile equilibrium, which to topple over it requires an amount of energy minute in comparison to that developed by its fall.

Evidently then if only that initial energy, small in amount but peculiar in quality, could be produced, this life preserving circulation of water could be controlled.

---//---

In endeavouring to solve this al-important problem I have discovered that the condensation of aqueous vapour is or may be brought about, by certain electrical discharges or oscillations of transcending intensity, which produce sudden and excessive compressions and rarefactions in the atmosphere.

Basing myself on experimental facts and observations, I explain their action as follows: The air being compressed is heated above the temperature of the surrounding, and a portion of the heat thus evolved is instantly radiated away.

Rarefaction succeeding, the air is cooled, but only slightly, as some of the vapour in it condenses, this being attended by a liberation of latent heat and partial dissipation of the same, by radiation or otherwise. Upon the air being again compressed, the droplets formed are not re-evaporated, as this would

---
20 As this text is difficult to obtain, I'll include a copy of the full text in the appendices.

require much more heat energy than consumed in their condensation, hence some of the heat generated is abstracted from the air as before. In the expansion following, the atmospheric gases are again cooled and more of the vapour is condensed. These actions are repeated in rapid succession, their effect being cumulative, resulting in a more or less copious precipitation of moisture. The energy used up in the compression and expansion of the air bears to the gravitational energy of the vapour condensed a relation not unlike that which exists between the energy required to pull the trigger of a gun and that stored in potential form, in the explosive. Obviously the trigger-mechanism may be more or less sensitive, so too the energy necessary to precipitate water and the quantity of the latter will vary greatly according to circumstances, admitting the correctness of the preceding explanation, the atmosphere subjected to sudden compressions and rarefactions will not be able to contain as much vapour as under normal conditions, and from this it might be inferred, that condensation will always result. But it must be borne in mind that the air is receiving continuously new heat from the surrounding media, which tends to counteract the effect of the discharges or oscillations. Although the direct rays of the sun pass quite freely through the vapour-charged air, the rays reflected from the earth's surface are almost wholly absorbed. To be effective, the energy of the oscillations must, therefore, be supplied at an enormous rate, even though for a very short time.

1917, May 18th

Minutes of the Annual Meeting of the American Institute of Electrical Engineers, Held at the Engineering Societies Building, Friday Evening

One day as I was walking in the forest a storm gathered and I ran under a tree for shelter. The air was very heavy, and all at once there was a lightning flash, and immediately after a torrent of rain fell. That gave me the first idea. I realized that the sun was lifting the water vapour, and wind swept it over

> the regions where it accumulated and reached a condition when it was easily condensed and fell to earth again. This life-sustaining stream of water was entirely maintained by sun power, and lightning, or some other agency of this kind, simply came in a trigger-mechanism to release the energy at the proper moment. I started out and attacked the problem of constructing a machine which would enable us to precipitate this water whenever and wherever desired.

We have already seen this next one in Tesla's autobiography. It was part of chapter 5 dealing with the TMT and this chapter started with a story about an avalanche, a small trigger creating a huge effect. This repeated story about lightning is very similar in that respect.

1919

Autobiography

> *One day, as I was roaming in the mountains, I sought shelter from an approaching storm. The sky became overhung with heavy clouds but somehow the rain was delayed until, all of a sudden, there was a lightning flash and a few moments after a deluge. This observation set me thinking. It was manifest that the two phenomena were closely related, as cause and effect, and a little reflection led me to the conclusion that the electrical energy involved in the precipitation of the water was inconsiderable, the function of lightning being much like that of a sensitive trigger.*

> *Here was a stupendous possibility of achievement. If we could produce electric effects of the required quality, this whole planet and the conditions of existence on it could be transformed. The sun raises the water of the oceans and winds drive it to distant regions where it remains in a state of most delicate balance. If it were in our power to upset it when and wherever desired, this mighty life-sustaining stream could be at will controlled. We could irrigate arid deserts, create lakes and rivers and provide motive power in unlimited amounts. This would be the most efficient way of harnessing the sun to the uses of man. The consummation depended on our ability to develop electric forces of the order of those in nature.*

1931, December

Our Future Motive Power

The sun raises the water to a height where it remains in a state of delicate suspension until a disturbance, of relatively insignificant energy, causes condensation at a place where the balance is most easily disturbed. The action, once started, spreads like a conflagration for a vacuum is formed and the air rushing in, being cooled by expansion, enhances further condensation in the surrounding masses of the cloud. All life on the globe is absolutely dependent on this gigantic trigger mechanism of nature and my extended observations have shown that the complex effects of lightning are, in most cases, the chief controlling agents. This theory, formulated by me in 1892, was borne out in some later experiments I made with artificial lightning bolts over 100 feet long, according to which it appears possible, by great power plants suitably distributed and operated at the proper times, to draw unlimited quantities of water from the oceans to the continents. The machines being driven by the waterfalls, all the work would be performed by the sun, while we would have merely to release the trigger.

This quote puts a date of 1892 on this idea, which falls in the period that Tesla was working on the development of his TMT. So both this quote, and the previous one, link this story to the TMT.

1933, December

Breaking Up Tornadoes

One on the greatest possible achievements of the human race would be the control of the precipitation of rain. The sun raises the waters of the ocean and winds carry them to distant regions, where they remain in a state of delicate suspension until a relatively feeble impulse causes them to fall to earth. The terrestrial mechanism operates much like an apparatus releasing great energy through a trigger or priming cap.

If man could perform this relatively trifling work, he could direct the life-giving stream of water wherever he pleased,

> *create lakes and rivers and transform the arid regions of the globe. Many means have been proposed to this end, but only one is operative. It is lightning, but of a certain kind.*

Most people take this story at face value and conclude that Tesla had plans to control rainfall. I believe this is only partly true. This is an example where Tesla tries to translate his electric discovery to another field. He noticed condensation of water vapour around his coils in Colorado Springs and thus thought of this analogy of the effect that he was trying to accomplish. Then, having discovered this effect he might as well take a patent on it. So he did, but he did not see it through.

Tesla discovered that our atmosphere contains huge amounts of energy received from the Sun (exactly as in this analogy) and we need to find some way to "condense" this energy into electric currents.

In his article "the Problem of Increasing Human Energy" he uses the analogy of "Burning Nitrogen" to describe the same process.

Conclusions of part 2

The Tesla code

In his writings after 1895 Tesla often resorts to analogies to describe his ideas and inventions. It is not always clear how these analogies should be interpreted but in some occasions he actually gives us the key and in some others we can get this key by comparing articles and time-lines. In his article "the Problem of Increasing Human Energy" Tesla explains large portions of his work using such analogies.

New source of energy

Tesla gives us plenty reasons to suspect that the Magnifying Transmitter would open up a new, clean source of energy. Looking for this in his "coded" work we find that this is almost certainly true.

As he describes it, his TMT cools a portion of "the medium" which then absorbs heat from its environment. This heat gets converted into an electric current which is consequently used to send power around the world. The TMT can then actually pick up this same power and continue to operate on the power that it has previously extracted from its environment.

Although this looks like a good description it is not immediately clear where this energy comes from and how this relates to our 2nd Law of Thermodynamics, which states that this would classify as a perpetuum mobile and would be impossible to actually work.

Cosmic rays

Tesla explains that he knows of this law but believes that this can not paint the complete picture. There must be something missing. In his own words, this describes the universe as a clockwork running down and would end in a cold and motionless universe.

As everything in the universe is rhythmical, goes up and down, this thing would be a unique exception and therefore unlikely to be complete. There has to be something that winds the clock up as it runs down.

Tesla believes he has found this force in what he calls "primary cosmic rays" (PCR). When something is submitted to a high voltage it tends to discharge to its environment. But what happens if its environment does not have anything to discharge to, or in other words, is a vacuum? It will then start to emit tiny electrified particles at enormous velocities, according to Tesla far exceeding the speed of light. These particle rays do not normally interact with matter, they simply pass through it. But when they do interact they cause radioactive decay and secondary cosmic radiation (SCR).

What we call "cosmic background radiation" and believe to be the remnants of the Big Bang, is what Tesla calls SCR and explains in an entirely different manner.

The PCR are in Tesla's view the ultimate source of energy. They wind up the "universal clockwork" as it runs down and therefore go by their very nature against the 2nd Law of Thermodynamics.

The question now remains; where are those very high voltages in our universe? Tesla's answer to that is that all stars including our Sun are under enormous electrical tensions and says to have experimentally established that our Sun is at an electric potential of 216,000,000,000 volts.

The idea that electricity plays an important role in our universe (astronomy) is gaining more and more traction in modern science. Many observed phenomena can be explained much better and simpler if we include electricity in our considerations.

Part 3: Experimental Evidence

Natural evidence

Nature provides our first evidence, though not complete it certainly adds to the likelihood that Tesla may have been correct.

From Wikipedia I get the following data for an average lightning strike.

Current: 30,000 Ampere
Charge: 15 Coulomb
Energy: 500,000,000 Joule

From the first two we can derive an average time interval of 0.5 ms.

Combining that with the energy gives us a power of 1 TW and a potential of 33 MV. If that potential was contained somewhere in the cloud this volume of air should have an electric capacitance of 0.45 µF, which would require a sphere with a radius of about 4 Km.

It is entirely unlikely that this charge spread over such a large non-conducting sphere can be funnelled into a lightning strike within 0.5 ms. Furthermore measurements have shown that a thundercloud recovers from a strike in less than 8 seconds, returning to its pre-strike condition. That would mean that the charge in this huge sphere gets replenished in such a short time.

Besides all these points we know that a lightning strike is preceded by a so called "stepped leader". That is a phenomenon in which we see a channel being formed in many steps of around 50 meters. Imagine a large high voltage region where a lot of electric charge is present, as pressurized air inside a balloon. Now when a needle is stuck in that balloon, do you expect this pressurized air to escape in many small steps or one big step?

One big step of course! Anything else would be ridiculous to assume. Now then, why would it be different for that high voltage region?

It is clear, I think, that there is a different process taking place here.

The measured current in the leader is in the order of 100 A, about 300 times less than in the "return strike". The individual steps lasts about 150 ns. That would imply the presence of about 15 µC, one millionth of that in the actual strike, thus reducing the above mentioned difficulties enormously.

Air at sea-level has a breakdown voltage of about 1MV per meter, so 33 MV can break through about 33 meters of air. If we go higher this distance gets longer. This matches the average step size pretty well. If Tesla is right then this discharge would collect additional charges from the air and cosmic rays which are consequently moved towards the end of the initial discharge, creating a new high voltage region. From there a new discharge will take place and this process will repeat. This matches our observations perfectly.

Also note that Tesla's requirements are satisfied: a slow build-up of a high voltage, followed by a sharp drop.

This is something that would happen in most multi-million volt discharges and we do indeed notice that these discharges can be significantly longer than the usual 1 meter per million volt, as can be seen in this picture.

FIGURE 1.2
Superlong negative discharge to a 110 kV transmission line wire. Voltage pulse amplitude, 5 MV. Courtesy of A. Gaivoronsky and A. Ovsyannikov, the Siberian Institute for Power Engineering.

If the stepped leader starts out with 15 µC of charge and builds up this ionized channel that delivers 1 million times that amount upon discharge, then surely there must have been charge added somewhere along the way.

Ring experiment

In this experiment I used my first 3-coil Tesla coil. The primary and secondary coils had a 1 m diameter, the primary was 1 turn, the secondary almost 36 turns. The extra coil had a 11.5 cm diameter and was about 1.9 m long, had 2,895 turns and a very small top-load.

In this experiment a copper ring was suspended horizontally around the top-load and next to this ring a ground wire.

The top-load discharged onto the copper ring, which in turn discharged to the ground wire.

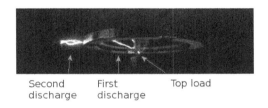

Second discharge   First discharge   Top load

In the pictures taken from this experiment we can see that the first discharge (top-load to copper ring) shows as a thin bluish thread, while the second discharge from the ring to the ground wire shows as a thick white discharge. This would seem to indicate a significantly higher current in the second discharge, in line with Tesla's theory.

Resonance measurement

For this experiment I only used the primary and secondary of the before mentioned Tesla coil, a small test coil that had the same resonance frequency and a small top-load connected with a long wire.

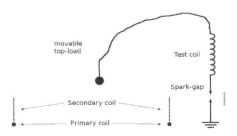

This movable top-load I could suspend anywhere near the primary and secondary. In the ground connection of the test coil there was a spark-gap and by measuring the maximum length of the spark I tried to get an impression of the electric field surrounding the primary and secondary.

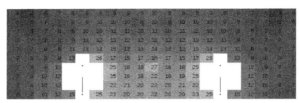

During these measurements I was called away and when I returned more than an hour later I found that I initially got significantly smaller sparks. Over the course of about 30 minutes these sparks grew until they were as long as they were before I was called away. This appeared to be a repeatable phenomenon; if I paused the experiment for a while then on restarting it the sparks were smaller and slowly grew. The time needed to minimize the spark length was about the same as the time needed to return to their maximum length, roughly 30 minutes.

This effect could be explained with the before mentioned interpretation of Tesla's patent 1,113,716; Fountain. The Tesla coil pulls up electricity from the Earth and this makes it easier to convey energy to the test coil.

IR thermometer

When taking measurements with an IR-thermometer around a top-load of a Tesla coil we notice an interesting phenomenon. When the coil is switched on we see a short spike in the temperature and when it is switched off we see a short lasting drop in the measured value.

When we shield the thermometer with grounded aluminium tape, this effect does get somewhat less. But the effect clearly remains and is very consistent in that it goes up when the coil is switched on and down when it is switched off.

I admit that there may be other forces at work here, but the consistency and the fact that it could be explained with Tesla's theory make it worth mentioning, I think.

Single terminal tube

Tesla said that the PCR are responsible for radioactive decay and that these can be produced with his single terminal vacuum tube.

For this test I used Tesla's specifications of such a tube given in patents 685,957 and 685,958 and a detector of my own design.

This detector consisted of a small tritium filled tube with a fluorescent coating that gives off a little bit of light because of the radioactive decay of the tritium. A photo-transistor was used to measure this light which was an indication of the speed at which the radio-active decay was taking place.

This detector was properly shielded and placed in front of the vacuum tube as the latter was connected to a source of about 1 MV.

The detector clearly responded to switching on the coils that supplied the 1 MV to the tube. But it did so everywhere near the coil even when the vacuum tube was removed.

So although this initially seemed to confirm Tesla's theories there was clearly something else going on. This may have something to do with the fact that I suffered many computer hardware failures while I did these experiments in my living room (before I had my lab).

It should also be noted that Tesla used much higher voltages then I could generate at that time.

Earth resonance

In this experiment I had a new coil in my new lab. The primary and secondary coils were 1.22 m in diameter, primary 1 turn, secondary 47 and a bit. The extra coil was 42.5 cm in diameter, 135 cm long and had 270 turns. I used ceramic capacitors (doorknob type) that worked very well, much better than the usual polypropylene ones.

The rotary spark-gap was driven by a stepper-motor that was controlled by an oscillator. Thus the rotational speed could be set with great precision and constancy.

It was immediately clear that the coil worked better when the spark-gap frequency was set to a multiple of 11.77 Hz.

The spark-gap was monitored together with the voltage on the primary capacitors and this showed that in some cases (when operated on the right frequency) the spark-gap fired even though the voltage was insufficient. I attribute this to the coils picking up an echo of an earlier pulse which proofs that the impulse travels through the Earth and returns in about 85^{21} ms as stated by Tesla.

Although strictly speaking I did not achieve resonance, at that time I did not have enough power available, this does prove the possibility thereof.

Apart from my own experiments on this subject, the facts reported by third parties (eye-witnesses) about Tesla's experiments simply can not be explained if Earth resonance were impossible as modern scientists believe.

Charge increase

In this experiment I measure the current through the ground wire of the secondary coil and that of the roof of the lab. The same coils as in the previous experiment were used only this time with polypropylene capacitors as my ceramic set was blown.

The coils were operated on half the capacity of the power supply as we experienced some issues at full power.

Though the result was not constant it did clearly show an increase in current on some occasions. I believe the reason to be that we did not always reach a sufficiently high voltage and also that the extra coil did not always resonate long enough (these two things are indeed related).

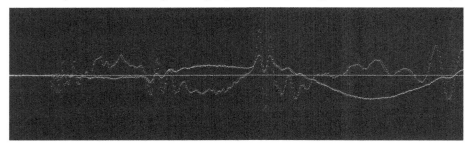

The lighter trace shows the current in the secondary, the darker that of the roof. It is obvious that the latter shows more activity and more current.

---

21   84.9 ms was measured

The results of this experiment contain a number of things that require further examination but they do support Tesla's claim.

Although the energy *we* get in our "free system" is less than we need to invest in our "exciting system", lightning already proves that it can be made profitable. It is just a matter of engineering skills, patience and money.

Conclusions of part 3

True or not?

Nature provides in lightning the most persuasive evidence that Tesla is most likely right. Tesla frequently refers to lightning as a source of his inspiration and it is obvious that he was trying to recreate lightning for some reason.

Either because he had lost his marbles or he was really onto something that he considered very important.

All evidence points in the direction of the latter option.

All experimental evidence that I have been able to obtain with the generous financial help of the investors in Ethergy Co. Ltd. points in that same direction.

In my mind it is extremely likely that Tesla's system would have worked and would have opened up a new, unlimited source of renewable energy. It is a shame, no, worse, it is a crime against humanity that billions of dollars are poured into the military and researches that have never produced anything useful while not one single government dollar has been spent on researching Tesla's legacy.

I am eternally grateful to the investors in Ethergy and to quote Tesla one more time:

> *Some day, I hope, a beautiful industrial butterfly will come out of the dusty and shrivelled chrysalis.*[22]

---

22 From "The Problem of Increasing Human Energy", June 1900, Century Illustrated Magazine.

Appendices

Patent application 213,055

June 17, 1904,

Production and Application of Electrical Force

TO ALL WHOM IT MAY CONCERN:

Be it known that I, Nikola Tesla, a citizen of the United States residing at New York, have made certain new and valuable discoveries and improvements in the production and application of electrical force, of which the following is a specification, reference being had to the accompanying drawing, forming part of the same.

It is well known that through the sun's heat water is evaporated and elevated to more or less considerable heights, where it remains in a state of delicate suspension, until disturbing influences arise which cause a condensation of vapour particles and their fall to the earth's surface under the action of gravity. Thus all vegetation and animal life on this globe is sustained. The body of water maintained in the air is like a heavy weight in labile equilibrium, which to topple over it requires an amount of energy minute in comparison to that developed by its fall.

Evidently then if only that initial energy, small in amount but peculiar in quality, could be produced, this life preserving circulation of water could be controlled. But though this truth may have been long ago recognized nothing, so far as I am aware, has been developed which would offer even the slightest possibility for man to gain mastery of this process of nature.

In endeavouring to solve this al-important problem I have discovered that the condensation of aqueous vapour is or may be brought about, by certain electrical discharges or oscillations of transcending intensity, which produce sudden and excessive compressions and rarefactions in the atmosphere.

Basing myself on experimental facts and observations, I explain their action as follows: The air being compressed is heated above the temperature of the surrounding, and a portion of the heat thus evolved is instantly radiated away.

Rarefaction succeeding, the air is cooled, but only slightly, as some of the vapour in it condenses, this being attended by a liberation of latent heat and partial dissipation of the same, by radiation or otherwise. Upon the air being again compressed, the droplets formed are not re-evaporated, as this would require much more heat energy than consumed in their condensation, hence some of the heat generated is abstracted from the air as before. In the expansion following, the atmospheric gases are again cooled and more of the vapour is condensed. These actions are repeated in rapid succession, their effect being cumulative, resulting in a more or less copious precipitation of moisture. The energy used up in the compression and expansion of the air bears to the gravitational energy of the vapour condensed a relation not unlike that which exists between the energy required to pull the trigger of a gun and that stored in potential form, in the explosive. Obviously the trigger-mechanism may be more or less sensitive, so too the energy necessary to precipitate water and the quantity of the latter will vary greatly according to circumstances, admitting the correctness of the preceding explanation, the atmosphere subjected to sudden compressions and rarefactions will not be able to contain as much vapour as under normal conditions, and from this it might be inferred, that condensation will always result. But it must be borne in mind that the air is receiving continuously new heat from the surrounding media, which tends to counteract the effect of the discharges or oscillations. Although the direct rays of the sun pass quite freely through the vapour-charged air, the rays reflected from the earth's surface are almost wholly absorbed. To be effective, the energy of the oscillations must, therefore, be supplied at an enormous rate, even though for a very short time. In the lightning discharges nature has provided an ideal means for precipitating moisture, and if electrical forces of such character could be artificially produced, an incalculable benefit to mankind could be secured through their intelligent control. Up to now such results have been thought unrealisable by any human agencies. Attempts to cause a fall of rain by firing bombs and guns have proved unsuccessful chiefly, I believe, on account of the limited activities thus obtainable amounting at most to a few thousand horsepower. But by gradual improvements of an electrical oscillator of the kind described in my patents Nrs: 645,576 and 649,621 which I have used in the transmission of energy without wires, I have have found it practicable to create in the atmosphere commotions, not only equalling, but by far surpassing in intensity those produced by lightning. Rates of many millions of horsepower are attainable with plants of very moderate size, by such means. I have thus made it possible to precipitate, at will, the vapour produced by the sun's heat and thus thus draw unlimited quantities of water to any locality desired, for irrigation, power supply or other useful purpose.

In the drawings Fig. 1 represents schematically two resonating circuits such as described in my patents before mentioned and applications pending, more especially in my application serial number 90,245[23]. These circuits, as will now be understood, produce electrical oscillations of transcending intensity. Adjustable terminals T T' may be provided between which the discharges may be made to pass in desired intervals, though this is not absolutely necessary, as the silent waves caused by the rapid and excessive variations of potential are likewise effective. It may be therefore sufficient to employ but one circuit such as illustrated.

Fig. 2 shows an arrangement for utilizing the fall of water precipitated by such apparatus, from a high reservoir R to a lower level for the purpose of driving dynamos as in a powerhouse P. The currents generated may be used in the production of the oscillations.

I claim as my invention:

1. *The method of precipitating moisture, which consists in producing in the atmosphere electric discharges or oscillations of sufficiently high potential. To cause a condensation of the aqueous vapour contained therein, as set forth.*
2. *The method of abstracting water from the atmosphere which consists in producing electrical discharges or oscillations of sufficient intensity, to cause the precipitation of the moisture therefrom, as set forth.*
3. *The method of precipitating moisture which consists in generating electrical impulses or oscillations of high potential*

---

23 Patent 1,119,732 Apparatus for Transmitting Electrical Energy

and causing thereby in the atmosphere sudden alternate compressions and expansions sufficient to effect a condensation of the aqueous vapour held in suspension therein, as set forth.
4. *The method herein described of precipitating moisture by electrical discharges or oscillations, and utilizing the energy of the precipitated water in its passage from a high to a low level, as set forth.*
5. *The method herein described of precipitating moisture by electrical discharges or oscillations, utilizing the energy of the precipitated water in its passage from a high to a low level, and using part of that energy in the production of the discharges or oscillations, as set forth.*

In testimony thereof...
N.T.

Filed June 16 work 3.6
Marcellus Bailey
501 F street NW.

Tesla's dynamic theory of gravity

If Tesla was right and a gaseous medium as he describes does exist then it is the presence of this medium that determines the value of ε_0 "the electric permittivity of the vacuum".

Electric permittivity is a property, and a property can not exist on its own. It needs a "carrier". For example if I have a red cube, I can not separate this into a red nothing and a colourless cube. Likewise electric permittivity needs a carrier. It simply can not exist on its own.

Now to say that the vacuum is this carrier is absurd as the word vacuum implies the absence of carriers. The vacuum can not have a colour, a weight, nor any property for that reason.

It was Einstein who in his lecture on May 5th, 1920 made this same observation though in different terms, and concluded that therefore an ether *must* exist. It is surprising to see that the followers of Einstein so strongly object to the concept of an ether.

Following Tesla's reasoning, the ether is probably not the carrier for this property, but his "medium" is. As this is a compressible medium it follows that ε_0 is not a universal constant but depending on the pressure of this medium.

As this medium "clings to matter" it follows that there is more of it where there is more matter, causing variations in ε_0.

A Russian (I believe) by the name of Vesselin Petkov has published a paper[24] describing how such variations can explain both inertia and gravity. I think this aligns perfectly with Tesla's work and could very well be the theory that he had in mind.

---

24 See https://arxiv.org/pdf/physics/9909019.pdf

Capturing cosmic rays

I separated this from the rest of this work as it is largely my speculation (as is the previous appendix) and I feel there is insufficient material in Tesla's own writings to make a definite statement on these subjects.

The question remains "how do we capture the energy of the cosmic rays?".

I believe it should go something like this:

We cool the ether down to near zero Kelvin. At this low temperature it becomes more susceptible to external influences.

The primary cosmic rays are tiny particles that shoot through this fluid medium and create whirls, vortices.

This vortex can be seen as a small rotating cylinder. When observed from one side it rotates clockwise, while seen from the other side it rotates counter-clockwise. Perhaps clockwise would mean a positive charge and counter-clockwise a negative, when ending on matter.

It is easy to see from this that positive and negative charges attract and annihilate each-other, forming a little bit longer cylinder, whereas like charges would repel.

If such a vortex would connect to an atom it's positive side would connect to the atom creating a positive ion, while the other end would appear to be a negative charge, or electron.

This situation would quickly return to its normal state if it weren't for the fact that this happens in an electric current which quickly separates the charges further and further.

Then if we collect these charges we effectively capture the energy of the cosmic rays.

This is of course pure speculation from my part, but it is the best that I can do based on Tesla's work.

--- area reserved for notes ---

--- area reserved for notes ---

Help bring back the Magnifying Transmitter

With the purchase of this book you are supporting my research and increasing my hopes that one day the world will see a full-blown Magnifying Transmitter.

If you have not bought this book or you wish to support my project even more, please feel free to make a donation in one of the cryptocurrencies below. Help create a better world!

BTC

LTC

Dash

Doge

THANK YOU!

Made in the USA
Monee, IL
27 June 2023